MOLECULAR BIOLOGY OF ERYTHROPOIESIS

ADVANCES IN EXPERIMENTAL MEDICINE AND BIOLOGY

A Continuation Order Plan is available for this series. A continuation order will bring delivery of each new volume immediately upon publication. Volumes are billed only upon actual shipment. For further information please contact the publisher.

MOLECULAR BIOLOGY OF ERYTHROPOIESIS

Edited by

Joao L. Ascensao

University of Connecticut Health Center
Farmington, Connecticut

Esmail D. Zanjani

Veterans Administration Medical Center
Reno, Nevada

Mehdi Tavassoli

Veterans Administration Hospital
Jackson, Mississippi

Alan S. Levine

National Heart, Lung, and Blood Institute
Bethesda, Maryland

and

F. Roy MacKintosh

University of Nevada
Reno, Nevada

PLENUM PRESS • NEW YORK AND LONDON

Library of Congress Cataloging-in-Publication Data

Symposium on Molecular Biology of Hemopoiesis (4th : 1988 : Reno,
  Nev.)
    Molecular biology of erythropoiesis / edited by Joao L. Ascensao
  ... [et al.].
        p.   cm. -- (Advances in experimental medicine and biology ; v.
  271)
    "Proceedings of the Fourth Annual Symposium on Molecular Biology
  of Hemopoiesis, held November 1-2, 1988 in Reno, Nevada"--T.p.
  verso.
    Includes bibliographical references.
    Includes index.
    ISBN-13: 978-1-4612-7897-9   e-ISBN-13: 978-1-4613-0623-8
    DOI: 10.1007/978-1-4613-0623-8
    1. Erythropoiesis--Molecular aspects--Congresses.
  2. Erythropoietin--Congresses.   I. Title. II. Series.
    [DNLM: 1. Erythropoiesis--congresses. 2. Molecular Biology-
  -congresses.   W1 AD559 v. 271 / WH 150 S9895m 1988]
  QP92.S95   1988
  612.1'11--dc20
  DNLM/DLC
  for Library of Congress   90-7255
                                                                        CIP

Proceedings of the Fourth Annual Symposium on
Molecular Biology of Hemopoiesis,
held November 1-2, 1988, in Reno, Nevada

Softcover reprint of the hardcover 1st edition 1989

A Division of Plenum Publishing Corporation
233 Spring Street, New York, N.Y. 10013

A PERSONAL TRIBUTE: RICHARD D. LEVERE, M.D.

Writing a personal tribute to my friend Richard Levere is a pleasurable task, though to do so within a constraint of space and without drawing upon most of the laudatory adjectives in the Oxford English Dictionary is a bit unsatisfying.

Richard Levere has been my friend for more than a quarter of a century. I marvel at that simple fact, and in its recalling, treasure it. Unalloyed and enduring friendship is not easy to find, and perhaps may be more diffi- cult to achieve between scientists who work in fields which so overlap each other as do his and mine. One could suppose that we are friends because we are both students of that wonderful man, the late Sam Granick, or because, despite our relative ages, I am really also a student of Richard Levere's. I left steroid endocrinology to enter the field of heme biology on the shaky assumption that because I knew something about the biochemistry of 4-ring steroid structures, I would necessarily know something about the biochemistry of other 4-ring structures such as heme; Richard Levere patiently taught me otherwise. I think perhaps, the most important reason that we are friends is simply that Richard chose, for reasons known only to a man with a good heart, to gift me with his friendship, and happily I was fortunate enough to recognize the value of the gift he was offering me. Those here who have also shared that gift know that the term, friend, is an unconditional one for Richard--a permanent state of grace in which the certainties implicit in friendship can be enjoyed without concern as to their circumstances.

It is inevitable that a man who can bestow such a quality of friendship on others should, when drawn to medicine as a career, become a splendid Physician--a word properly capitalized to affirm its significance as a formal title to which we impute all of the personal and professional char- acteristics of those who best represent the healing art. Richard Levere is indeed a splendid physician--a patient, compassionate, wise counselor; a diagnostition of the first rank; a therapist of consumate skill. If medicine as a whole had more physicians in its ranks as dedicated, competent and humane as is Richard Levere, many of the ills of our profession would resolve themselves.

It can be no surprise that an academic leader of Richard's calibre and temperament would enrich the scholarly ambience of any institution with which he is associated. That has proved to be the case in each of the two medical schools with which he has been principally affiliated during his career. And that each has benefitted enormously from his dedi- cation to superior medical practice and medical science is evident to all those who are familiar with the academic scene in New York. It is an aspiration common to all of us that we might become superb clinicians, fine teachers, and excellent scientists; but that is a hope more often

felt in the heart than is manifest by our labours. Richard Levere belongs
to that very small group of my contemporaries who have managed to make
that aspiration a reality throughout their professional lives.

This sort of text is not the place for a detailed review of the scientific achievements of one's friends. But I think it is important to note
that Richard Levere's publications recapitulate the man in a remarkable
way. Clinical reports in the New England Journal of Medicine, and Blood
intermingle with biochemical studies in the Proceedings of the National
Academy of Sciences and the Journal of Biological Chemistry; cytochemical
observations in the Journal of Cell Biology alternate with molecular genetic
investigations in Biochemical and Biophysical Research Communications;
work on heme and globin synthesis in the Journal of Experimental Medicine
moves on to the intricacies of arachidonic acid metabolism in Science.

I believe I know what this all means. Richard Levere is the medical
equivalent of a modern day Jeffersonian man. I do not know what committees
of the Continental Congress Thomas Jefferson served on, or if he ever
chaired a symposium, as Richard is doing at this meeting. But I am confident
that if there has been a occasion for them to meet, there would have been
a strong sense of affinity and much animated conversation between them
and they would have parted as good friends.

Attallah Kappas, M.D.
The Rockefeller University
New York, New York

PREFACE

This volume presents the proceedings of the Fourth Annual Symposium on the Molecular Biology of Hemopoiesis, held in Reno, Nevada, November 1 and 2, 1988. Its focus on erythropoiesis represents an attempt to cover a rapidly expanding field, which has gone from elegant studies of erythropoietin physiology, to molecular biology, to clinical applications and again to physiology. The rapid development has been made possible by cloning of erythropoietin gene and the availability of recombinant hormone.

The regulation of heme and its derivatives has also been aided by techniques of molecular biology; there is now a concerted effort to better understand how these enzymes contribute to proliferation, differentiation and maturation of the erythron. Globin gene derrangements have been targets of recent research in an attempt to correct the defect by genetic engineering. In the chapters of this book, several groups "expressed" their views on this subject. Finally, we analyze various regulators of erythropoiesis, both in vivo and in vitro.

Dr. Richard Levere was a pioneer in many studies of heme metabolism and of erythropoiesis. He has been a generous supporter of research in this field and of our past meetings. It is only fitting that this volume should be dedicated to him.

The Editors

ACKNOWLEDGEMENTS

The organizers are indebted to the following organizations for their partial support of this meeting: Research Section, V.A. Medical Center, Reno, NV.; Department of Medicine, New York Medical College, Valhalla, NY; Division of Hematology-Oncology, Department of Medicine, University of Nevada, Reno, NV.; Amgen, Inc, Thousand Oaks, CA; Ortho-Biotech, Raritan, NJ; Chugai-Upjohn, Inc., Kalamazoo, Mi.; Ms. Sheila Rich provided invaluable assistance and Ms. Melanie Yelity (Plenum Publishing Corp.) was always available.

CONTENTS

BIOLOGY OF ERYTHROPOIETIN

ERYTHROPOIETIN: FROM MOUNTAIN TOP TO BEDSIDE

A. J. Erslev and J. Caro

Jefferson Medical College of Thomas Jefferson University
Department of Medicine
Philadelphia, PA 19107

Erythropoietin is a growth factor for erythroid and myeloid progenitor cells and a differentiation hormone necessary for the blast transformation of CFU-E to proerythroblasts.

The existence of such a regulating factor or hormone was dimly perceived in the nineteenth century in order to explain the maintenance of an "interior milieu" with an optimal number of red cells in the circulation. In the 1850's, Dennis Jourdanet observed that people living at high altitude had an increase in their red cell count and he and his friend, Paul Bert, realized that an increased red cell count would provide a survival advantage for people living at a low oxygen pressure[1]. That the low oxygen pressure actually initiated an increase in red cell production was not, however, realized until much later and it took almost a hundred years to explain how hypoxia affects red cell production. In 1950, Reissmann provided indirect evidence for the existence of a factor released by hypoxia and capable of stimulating red cell production in the bone marrow[2] and in 1953, Erslev showed directly that such a factor is present in plasma of anemic and hypoxic animals[3], and it was named erythropoietin.

A few years later, studies by Jacobsen and co-workers showed that the factor was produced in the kidney and that the kidney actually worked as an endocrine gland[4]. Attempts to localize the site of production in the kidney by crude methods suggested that it was made not by the glomeruli but by tubular or interstitial cells[5]. Recent studies using probes for erythropoietin messenger RNA have pinpointed the site of production to interstitial renal cells, possibly endothelial cells lying in the immediate proximity of the proximal tubules[6,7]. The location here suggests that the tubular cells may in some way provide oxygen sensing and more recent work indicates the existence of a triggering heme protein[8]. Presumably hypoxia will change the confirmation of this heme protein resulting in the release of a short range signal capable of activating the erythropoietin gene.

The human erythropoietin gene has been pinpointed to chromosome 7 and shown to express considerable homology with erythropoietin genes of mice and monkeys[9]. It codes for a polypeptide consisting of 165 amino acids with a molecular weight of 18,000 dalton. This polypeptide backbone is extensively glycosylated in the Golgi apparatus and the final product, erythropoietin, has a molecular weight of about 34,000 dalton. Its half--life is about 6 hours but its site of degradation is still not known. In addition to the interstitial cells in the kidney, hepatocytes and

Molecular Biology of Erythropoiesis
Edited by J. L. Ascensao *et al.*
Plenum Press, New York

possibly Kupffer cells have been shown to contribute 10-15% of total circulating erythropoietin[10]. The hepatic contribution may be a remnant from fetal life during which the liver appears to be the sole producer of erythropoietin with the kidneys first taking over at the time of birth[11].

In the bone marrow, erythropoietin acts primarily as a differentiation hormone and it appears to be necessary for the transformation of the most mature progenitor cell, CFU-E, to a proerythroblast[12]. In the absence of erythropoietin, no such transformation occurs and the CFU-E will die. The more erythropoietin present the more CFU-E's will be transformed, each of them producing a progeny of about 16 red blood cells (see Figure 1). Erythropoietin has also a less well defined capacity to act as a growth factor on all progenitor cells as well as on the most immature of the erythroblasts. The growth of the progenitor cells, however, is more dependent on the presence of locally produced growth factors such as IL-3 or GM-CSF appropriately termed burst promoting factors.

Of these actions only the differentiating function can be considered a true hormonal effect and be incorporated into a feedback system controlling red cell production (Figure 2).

The concentration of erythropoietin in the circulation can be measured by biologic or radioimmune assays[13]. The biologic assay measures the erythropoietic action either in vitro in a suspension of progenitor cells or in vivo using animals in which endogenous erythropoietin has been eliminated by induced polycythemia. The radioimmune assay is based on using pure erythropoietin and a polyclonal antibody responsive to epitopes on the erythropoietin molecule. So far the results of the assay have been interchangeable (Figure 3) and we have yet to identify the production

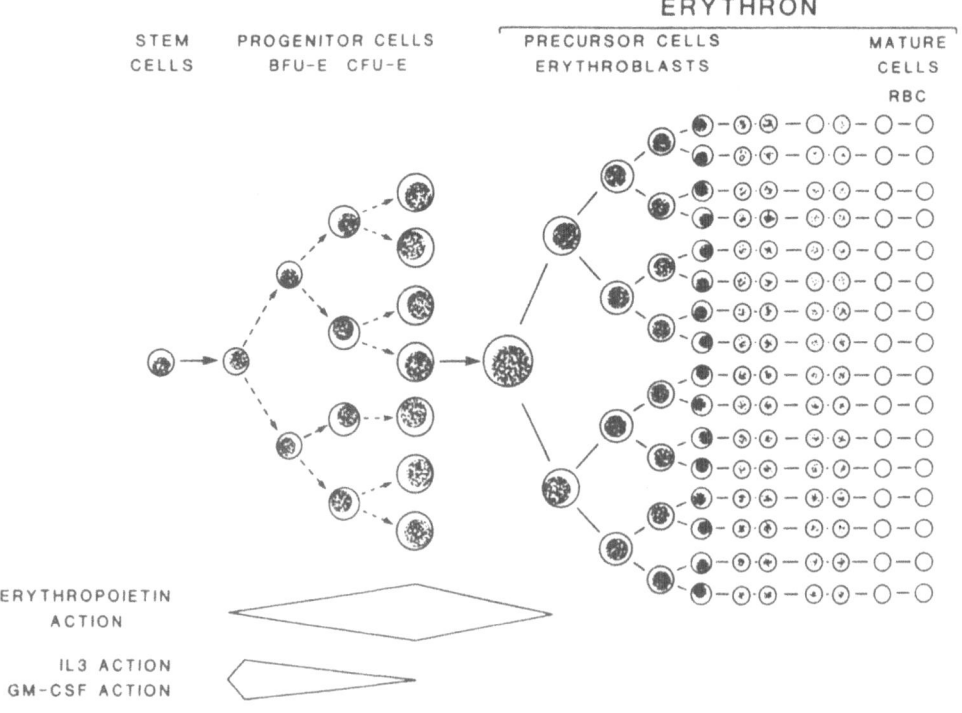

Fig. 1. Erythropoiesis. A schematic overview of the kinetics of red cell production.

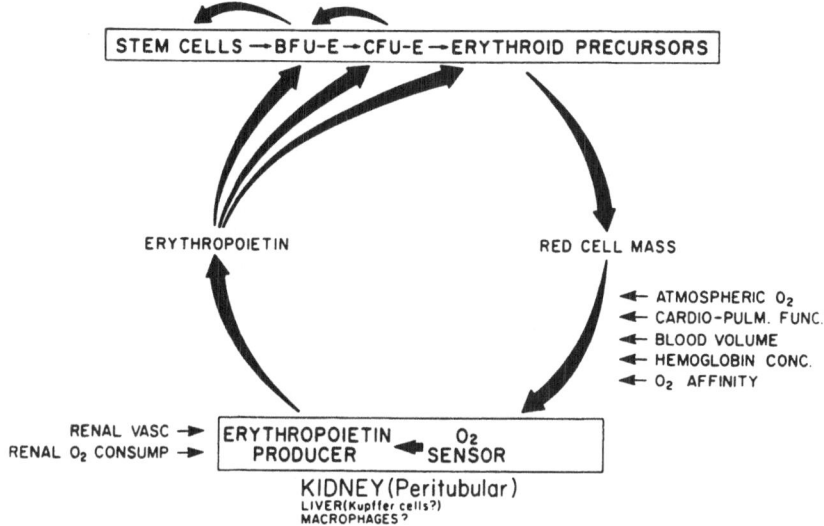

Fig. 2.   Bone Marrow.   The 1988 version of the feedback system
          which controls red cell production

of an immunologically active but biologically inactive hormone.

As can also be seen on Figure 3, there is a steep increase in erythropoietin concentration at low hemoglobin or hematocrit values, much steeper
than could be expected from in vitro dose response studies of progenitor
cells. It may be that the high concentrations of erythropoietin observed
in severe anemias are made primarily to recruit early progenitor cells
(BFU-E) rather than merely differentiate late progenitor cells (CFU-E)
to erythroblasts.

The measurements of erythropoietin titers can be of diagnostic and
therapeutic importance in both polycythemias and anemias.

In primary polycythemia or polycythemia vera, the bone marrow acts
autonomously and will overproduce red blood cells causing an increase
in hemoglobin and hematocrit. This increase will in turn produce renal
hyperoxia and cut off all erythropoietin production. Consequently, one
of the most consistent diagnostic features in polycythemia vera is a
near absence of erythropoietin production[14]. Since there are many other
features which help to provide a diagnosis of polycythemia vera, erythropoietin measurements are rarely needed. However, in difficult cases,
the diagnosis should be questioned if erythropoietin titers are higher
than 3-5 mU/ml (Figure 4).

In secondary polycythemia, on the other hand, tissue hypoxia will
generate erythropoietin which in turn stimulates the marrow to produce
more red blood cells. Consequently, erythropoietin titers are usually
elevated in contradistinction to those in polycythemia vera.

Polycythemia may also be caused by an inappropriate production of
erythropoietin by certain renal or extrarenal tumors or cysts. In the
absence of clinical features of either polycythemia vera or secondary
polycythemia, it may be of value to measure erythropoietin titers in
order to identify an erythropoietin producing lesion[15].

Since erythropoietin is produced primarily in the kidneys, it was

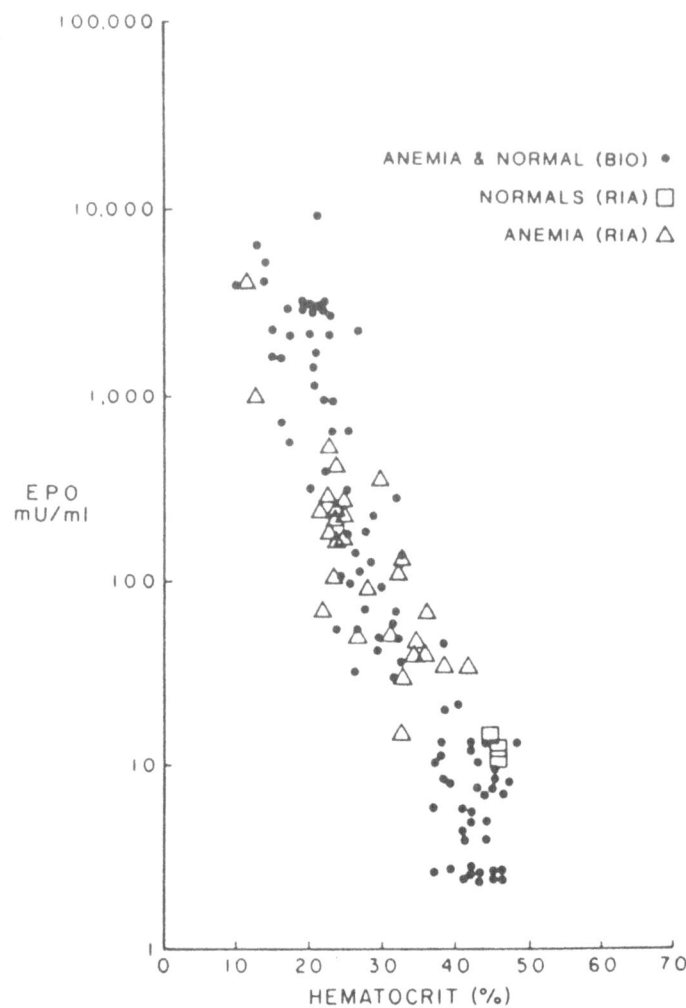

Fig. 3. Erythropoietin titers measured by in vivo bioassay or in vitro radioimmune assay and their
relation to the hematocrit of blood.

early realized that patients with chronic renal disease may have a deficient
production of erythropoietin. This hypothesis was confirmed by the assay
of erythropoietin in patients with chronic renal disease (Figure 5) and
it was realized that whenever erythropoietin became available for clinical
use, it could be of importance as a replacement agent for patients with
anemia of chronic renal disease. In 1985, the gene was isolated through
the marvels of molecular technology. It was inserted in hamster cells
which have the capacity to glycosylate polypeptides and mass production
of human erythropoietin began. The early clinical trials have showed
conclusively that erythropoietin will stimulate the bone marrow of patients
with chronic renal disease and in fact abolish the anemia[16,17]. Subsequent
studies have confirmed these findings and also have shown that an increase
in hematocrit may be associated not only with an improvement in the quality
of life but also occasionally in aggravation of hypertension and in production of clotted shunts. Except for prematurity[18], no other condition
has been associated with a significant and consistent decrease in erythropoietin production and replacement therapy with erythropoietin probably
will be restricted primarily to patients with chronic renal disease or
without kidneys.

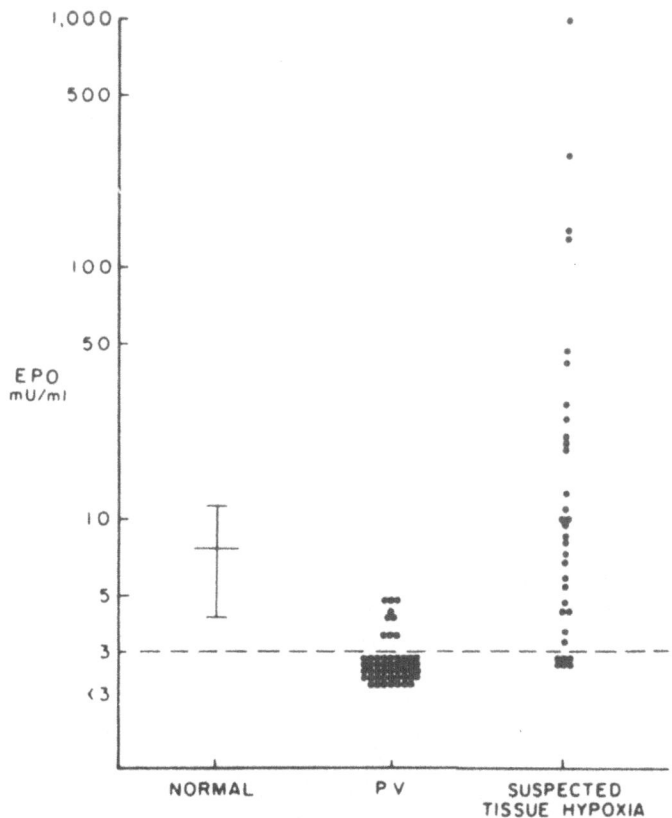

Fig. 4. Erythropoietin titers in blood from normal
individuals and patients with primary polycythemia vera or suspected secondary polycythemia.

However, pharmacologic therapy of refractory anemias may have a
place when unlimited amount of erythropoietin is available. In other
words, even if patients have normal kidneys producing adequate amount
of erythropoietin the bone marrow function may be improved by providing
additional exogenous erythropoietin. This is especially pertinent for
the anemia of chronic disease such as rheumatoid arthritis in which the
bone marrow appears to be somewhat resistant to appropriately produced
endogenous erythropoietin. So far only a few case reports have been
published but in these patients erythropoietin has restored hemoglobin
to normal and possibly improved their quality of life[19]. The same may
be true for hematologic malignancies with dysfunctioning marrows but
large amounts of exogenous erythropoietin will probably be needed in
order to make these patients transfusion independent. Clinical trials
are in progress but the results are still not known.

Due to its growth enhancing effect on progenitors and early erythroblasts erythropoietin will cause a rapid transit through the progenitor
and precursor department with the production of fetal hemoglobin containing
cells. This feature could be of importance in patients with sickle cell
anemia whose clinical problems would be alleviated by an increase in
fetal hemoglobin. Unfortunately, it does not appear that erythropoietin

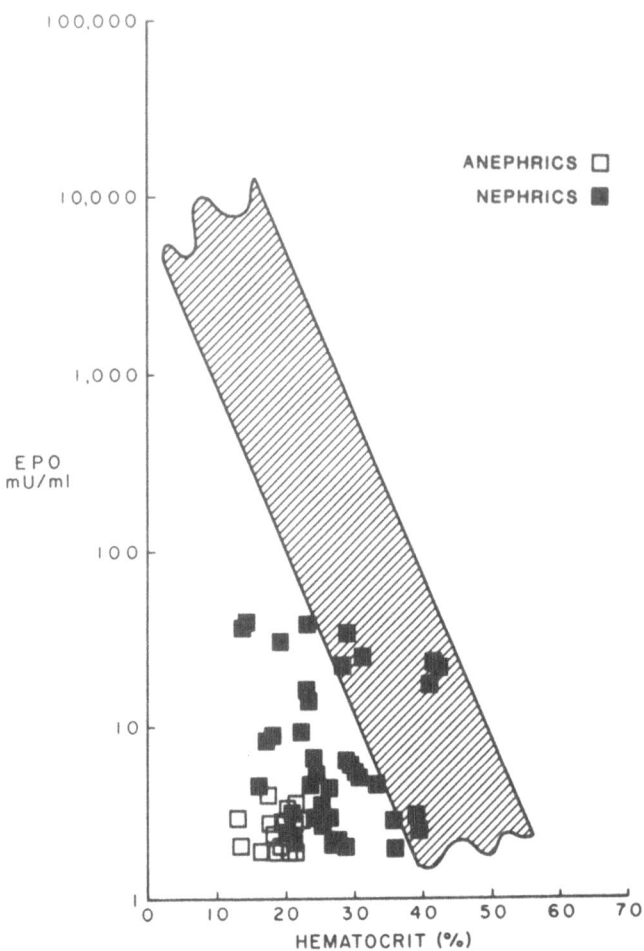

Fig. 5. Erythropoietin titers in blood from anephric
patients (hollow squares) and from patients
with severe renal failure (solid squares) and
their relation to the hematocrit of blood.
The cross-hatched area represents erythropoietin titers of anemic, but nonuremic
patients (see Fig. 3).

per se will produce that much extra fetal hemoglobin but it may supplement
the effect of other fetal hemoglobin producing agents such as hydroxyurea[20].

     Because of the inherent dangers of transfusion, more and more people
prefer to donate their own blood before an elective operation in order
to have their own blood on reserve if need should be. Such donations
could be accelerated if the patient simultaneously receive injections
of exogenous erythropoietin.

     All in all, erythropoietin has come a long way from its discovery
as an erythroid growth factor 40 years ago[21]. It is universally active
in stimulating the rate of red cell production and in patients with chronic
renal disease it has been shown to eliminate their anemia. It seems
likely that it also will ameliorate the anemia in many other patients
rendering them transfusion independent and improving their quality of
life.

REFERENCES

1. Erslev, A.J. 1981. Erythroid adaptation to altitude. Blood Cells 7: 495-508.
2. Reissmann, K.R. 1950. Studies on the mechanism of erythropoietic stimulation in parabiotic rats during hypoxia. Blood 5: 372.
3. Erslev, A.J. 1953. Humoral regulation of red cell production. Blood 8: 349.
4. Jacobson, L.O., E. Goldwasser, W. Freed, and L. Plzak. 1957. Role of the kidney in erythropoiesis. Nature 179: 633.
5. Caro, J., and A.J. Erslev. 1984. Biologic and immunologic erythropoietin in extracts from hypoxic whole rat kidneys and in their glomerular and tubular fractions. J. Lab. Clin. Med. 103: 922.
6. Koury, S.T., M. Boudurant, and M.J. Koury. 1988. Localization of erythropoietin synthesizing cells in murine kidneys by in situ hybridization. Blood 71: 524.
7. Lacombe, C., J-L DaSilva, and P. Bruneval, et al. 1988. Pertibular cells are the site of erythropoietin synthesis in the murine hypoxic kidney. J. Clin. Invest. 81: 620.
8. Goldberg, M.A., S.P. Dunning, and H.F. Bunn. 1988. Regulation of the erythropoietin gene: Evidence that the oxygen sensor is a heme protein. Science 242: 1424.
9. Jacobs, K., L. Shoemaker, and R. Rusersdorf, et al. 1985. Isolation and characterization of genomic and cDNA clones of human erythropoietin. Nature 313: 806.
10. Fried, W. 1972. The liver as a source of extrarenal erythropoietin production. Blood 40: 671.
11. Zanjani, E.D., J. Poster, and H. Burlington, et al. 1977. Liver as the primary site of erythropoietin formation in the fetus. J. Lab. Clin. Med. 89: 640.
12. Spivak, J.L. 1986. The mechanism of action of erythropoietin. Int. J. Cell Cloning 4: 139.
13. Erslev, A.J., J. Wilson, and J. Caro. 1987. Erythropoietin titers in anemic, non-uremic patients. J. Lab. Clin. Med. 109: 429.
14. Erslev, A.J. and J. Caro. 1983. Pathophysiology and classification of polycythemia. Scand. J. Haemat. 31: 287.
15. Erslev, A.J., and J. Caro. 1984. Pure erythrocytosis classified according to erythropoietin titers. Am. J. Med. 76: 57.
16. Winearls, C.G., M.J. Pippard, and M.R. Downing, et al. 1986. Effect of human erythropoietin derived from recombinant DNA on the anemia of patients maintained by chronic haemodialysis. Lancet II: 1175.
17. Eschbach, J.W., J.C. Egrie, and M.R. Downing, et al. 1987. Correction of the anemia of end stage renal disease with recombinant human erythropoietin. N. Engl. J. Med. 316: 71.
18. Rhondeau, S.M., R.D. Christensen, and M.P. Ross, et al. 1988. Responsiveness to recombinant human erythropoietin of marrow erythroid progenitors from infants with the anemia of prematurity. J. Ped. 112: 935.
19. Means, R.T., N.J. Olsen, and S.B. Krantz, et al. 1987. Treatment of the anemia of rheumatoid arthritis with recombinant human erythropoietin. Blood 70(Supp) 139A.
20. Al-Katti, A., T. Papayannopoulou, and G. Knitter, et al. Cooperative enhancement of F-cell formation in baboons treated with erythropoietin and hydroxyurea. Blood 72: 817.
21. Erslev, A.J. 1987. Erythropoietin coming of age. N. Engl. J. Med. 316: 101.

SERUM ERYTHROPOIETIN AND HEMOGLOBIN AFFINITY FOR OXYGEN IN

PATIENTS PHLEBOTOMIZED FOR POLYCYTHEMIA VERA

Ruth A. Cohen, Gisela Clemons,
and Shirley Ebbe

Lawrence Berkeley Laboratory
University of California
Berkeley, CA 94720

ABSTRACT

Serum erythropoietin (Ep) was measured by radioimmunoassay before
and 24 hours after therapeutic phlebotomies in patients with polycythemia
vera (PV) and in normal subjects before and after phlebotomies of comparable
volumes. In addition the in vivo oxygen affinity of hemoglobin (P_{50})
was calculated, and red cell indices and 2,3 DPG values were measured.
Paired t tests determined whether the differences between pre-and post--
phlebotomy values were statistically significant.

Blood hemoglobin (Hb) levels declined after phlebotomy, and generally
continued to be at or above normal levels. Serum Ep increased after
phlebotomy in both groups of subjects. The in vivo P_{50} value for patients
with PV (29.4± 0.4 mmHg) was significantly ($p < 0.005$) greater than the
normal value (27.2± 0.5).

Seven of the PV patients (5 males, 2 females) were restudied. Their
Hb and hematocrit values were either normal or slightly higher than normal.
The MCV for 4 males and 1 female was below normal. The MCHC was slightly
lower than normal and reticulocytosis was not present. The male PV patients
had greater than normal 2,3 DPG values and most had right-shifted P_{50}
values. There was no correlation between 2,3 DPG values and P_{50} values.
The female patients did not have 2,3 DPG values consistently greater
than normal and their P_{50} values were not right shifted.

These results showed that serum EP increased in response to small
reductions in Hb even when subnormal Hb values were not produced. The
reduced affinity of Hb in PV patients may explain earlier observations
that patients with PV have lower levels of urinary or plasma Ep than
normals with the same hematocrits.

INTRODUCTION

Erythropoietin (Ep) levels in human blood vary inversely with the
blood hemoglobin concentration (Hb) over a wide range from polycythemia
vera (PV) at one end of the scale to severe anemia at the other[1-7]. This
relationship suggests that blood Ep levels may be responsive to small

Molecular Biology of Erythropoiesis
Edited by J. L. Ascensao *et al.*
Plenum Press, New York

changes in Hb at all tolerable Hb levels. Adamson[8] found that patients with PV excreted less Ep in their urine than did normals with the same hematocrit (Hct). Garcia et al[6] reported that the average plasma Ep in a group of men with treated PV was lower than in a group of normal men, even though the average Hb levels were identical in the two groups. The subnormal Ep in patients suggested the possibility that the relationship between Ep levels and blood Hb may differ from normal in PV even though an inverse correlation between Ep and Hb may persist in PV[2,6].

In the present studies, Ep responses to small changes in Hb were determined in patients with PV, in whom Hb was at or above normal levels before and after therapeutic phlebotomies. In addition the in vivo hemoglobin oxygen affinity (P_{50}) and 2,3-Diphosphoglycerate (DPG) levels were determined in both male and female PV patients and normal controls.

Decreased hemoglobin oxygen affinity, right shifted P_{50}, can be a contributory factor to lower than normal Ep levels for a given hematocrit. We found that PV patients tended to have a lower hemoglobin affinity for oxygen and right shifted P_{50} values when compared with normal controls which could account for their decreased plasma Ep levels. Nevertheless, PV patients responded to therapeutic phlebotomy with increased plasma Ep levels.

MATERIALS AND METHODS

Subjects

Human subject involvement in these experiments was approved by the Lawrence Berkeley Laboratory Human Use Committee and the University of California, Berkeley, Committee for the Protection of Human Subjects. There were 12 Adults with PV and 8 normal adults. Patients had been treated for 5-19 years in the Donner Clinic at the time of study. Eight of the 12 patients fulfilled the Polycythemia Vera Study Group criteria for diagnosis of PV[9]: (a) Increased red cell mass + (b) normal arterial oxygen saturation + (c) splenomegaly or 2 of the following, thrombocytosis, leukocytosis and elevated leukocyte alkaline phosphatase. Arterial blood gases for the other 4 patients were not recorded, and the patients had no clinical evidence of hypoxia or cardiac or pulmonary disease. All 12 patients were on a program of therapeutic phlebotomies; 8 had also received radioactive phosphorus or chemotherapy.

Protocol

On day 0, phlebotomy of 490-500 cc (0.5-0.9% of body weight) was done, and the initial portion of the blood was collected for the tests listed below. On day 1, 11-25 cc of blood were drawn. The protocol was worked into the treatment schedule for each patient with their consent.

Blood Tests

Venous blood was drawn anaerobically for blood gas determination into heparinized, 5 cc glass syringes on day 0 and on day 1. The hemoglobin oxygen affinity, "in vivo" P_{50}, was calculated with the use of the Hill[(1)] equation assuming n=2.7[10]. The pO2 was measured on a pH/blood gas analyzer (model 813, Instrumentation Laboratory, Lexington MA) and the oxygen

(1) Hill Equation: $Y = 100 [P/P1/2)^n]$ (Y=%Hb combined with O_2 at pressure P, P1/2 is the oxygen pressures at which 50% of the Hb exists as oxyhemoglobin, n is constant)

saturation was determined spectrophotometrically (model 282, Cooximeter, Instrumentation Laboratory). P_{50} determinations were measured approximately one year apart in PV patients. Blood Hb was measured spectrophotometrically, red cell count, Hct and MCV were determined on a Coulter Counter and the MCHC was calculated. The % of reticulocytes was determined microscopically, and the Sigma diagnostic kit no. 35-UV was used to determine the concentration of 2,3-DPG.

Blood pH and pCO_2 values were also measured and did not differ significantly from normal control values. The caroxyhemoglobin (COHb) for all PV patients and controls was 1-2%

Erythropoietin

Serum Ep was measured in duplicate by radioimmunassay (RIA) as previously described[6]. All samples from a given individual were analyzed in the same assay. The RIA produced the same results on residual plasma samples obtained at the same time as serum samples.

Statistics

Pre- and Post-phlebotomy values were compared by paired t tests. Differences between groups were tested by the t statistic for 2 means.

RESULTS

Blood Hb levels and corresponding serum Ep levels are presented in Table 1 for samples obtained immediately before and 24 hours after phlebotomy from patients with PV and normals. The values represent 15 phlebotomies of 10 non-smoking PV patients (7 men, 3 women; 5 men were studied twice). The normal values represent serum Ep and Hb levels for 4 men and 3 women (all non-smoking) before and after phlebotomy. On day 0, prior to phlebotomy, Hb was significantly higher ($p < 0.025$) and Ep lower ($p < 0.001$) in PV patients than in normal control patients. For the PV group, Hb was less ($p < 0.001$) and Ep was greater ($p < 0.01$) 24 hours after phlebotomy when compared with the values obtained at the time of phlebotomy. Normals also showed a decrease in Hb ($p < 0.005$) and an increase in Ep ($p < 0.01$), 24 hours after phlebotomy. Neither group displayed a significant correlation between Hb and Ep before phlebotomy (PV patients, r=0.63; normals, r=0.45). After phlebotomy neither group displayed any correlation between Hb and Ep (PV, r=0.01; normals, r=0.22).

The calculated "in vivo" oxygen affinity of Hb (P_{50}) and corresponding hemoglobin values for 8 of the PV patients were determined before and after phlebotomy. The mean P(50) value ($\pm$)SEM for PV patients (19 measurements) before phlebotomy was 29.4$\pm$0.4; and, 24 hours after phlebotomy (8 measurements) was 29.9$\pm$1.0 as shown in Table 2. The pre- and post-phlebotomy P_{50} mean values for the PV group were significantly greater ($p < 0.005$) than the normal mean value, 27.2$\pm$0.5 (Table 2), which coincides with accepted normal values[11].

In order to confirm this finding and determine 2,3-DPG levels, 7 of the original PV patients were restudied. Blood Hb, Hct, MCV, MCHC, % reticulocytes, 2,3-DPG, and calculated "in vivo" P_{50} values are shown in Table 3 for the PV group and the normals. The Hb, Hct and % reticulocytes for all PV patients were either normal or just slightly higher than normal. However, the MCHC was lower and 2,3-DPG values were higher in the PV group when compared with the normal controls.

PV patients #1-5 showed the greatest increase in 2,3-DPG when compared

Table 1. Blood Hemoglobin and Serum Erythropoietin Before (day 0) and 24 Hours After (day 1) Phlebotomy

Polycythemia Vera

Subject #	Hemoglobin, gm/dL		Erythropoietin mU/ml	
	day 0	day 1	day 0	day 1
1	17.5	14.8	12.3	17.3
	16.8	15.1	9.6	15.8
2	18.8	18.6	13.8	13.8
	16.6	15.8	7.9	10.2
3	18.3	17.3	14.5	14.9
	15.8	15.3	10.3	13.3
4	17.5	15.3	14.7	16.8
	15.3	15.9	7.1	7.0
5	16.9	15.6	9.5	11.5
	16.6	15.5	9.7	8.7
6	16.8	14.8	11.0	12.4
7	15.2	14.2	10.0	9.5
8	16.7	15.2	16.4	18.7
9	16.0	15.5	10.1	10.2
10	16.5	15.1	12.6	14.2
MEAN	16.8	15.6	11.3	13.0
SEM	±1.0	±1.0	±2.6	±3.3

Normal

Subject #	Hemoglobin, gm/dL		Erythropoietin mU/ml	
	day 0	day 1	day 0	day 1
4	16.5	14.4	21.4	33.3
5	16.8	15.0	22.5	36.1
6	14.3	12.3	13.3	18.7
7	12.9	12.6	14.5	21.4
8	13.3	12.5	16.9	19.5
9	16.9	16.0	12.6	14.8
10	16.3	14.7	18.8	22.0
MEAN	15.3	13.9	17.1	23.7
SEM	±1.6	±1.3	±3.6	±7.3

Statistical analyses by paired t tests, probability that the two named values are the same:
1. Polycythemia vera, hemoglobin day 1 vs day 0: $p \leq 0.001$. (d1<d0)
2. Polycythemia vera, erythropoietin day 1 vs day 0: $p \leq 0.01$. (d1>d0)
3. Normal, Hemoglobin day 1 vs day 0: $p \leq 0.005$. (d1<d0)
4. Normal, erythropoietin day 1 vs day 0 $p \leq 0.01$. (d1>d0)

with normal values. Patients #1-5 were males; and, while most had right shifted P_{50} values, there was no correlation between the P_{50} and 2,3-DPG values. PV patients #6 and 10, both females, did not show consistently greater than normal 2,3-DPG values and their P_{50} values were not right shifted.

DISCUSSION

Serum Ep increased after phlebotomy even when the Hb did not fall below normal in the 10 non-smoking PV patients presented here. This conclusion is based on the statistical analysis of 15 phlebotomies in patients with PV and 7 normal controls. However, as individuals a decrease, an increase, or no change was measured in serum Ep levels. When Ep increased, the change was frequently small. Thus, while small changes in serum Ep may have physiological significance they may not achieve statistical significance in individuals or in small groups of subjects.

Table 2, Calculated in vivo P_{50} values and corresponding hemoglobin in non-smoking patients with polycythemia vera and normal controls

POLYCYTHEMIA VERA						NORMAL		
SUBJECT #	Months[a]	P_{50}[b]	Hb[c]	P_{50}[b']	Hb[c']	SUBJECT #	P_{50}[b]	Hb[c]
1	1	32.0	16.2	31.2	14.6	1	28.1	16.2
	4	28.1	16.7	–	–			
2	1	27.0	18.6	28.0	18.6	2	28.2	13.8
	4	29.0	17.9	–	–			
	2	29.9	18.4	–	–	3	24.7	14.9
3	2	30.8	18.0	29.2	17.3	4	27.0	15.6
	1	28.4	16.8	–	–		27.7	16.5
	1	29.3	15.6	–	–			
4	1	28.2	15.8	27.6	15.3	5	27.4	16.8
	2	30.6	14.4	–	–			
	2	29.2	15.3	–	–			
5	5	30.5	17.5	30.5	15.6			
	1	28.9	15.2	–	–			
	1	30.1	13.8	–	–			
6	5	30.8	16.4	27.7	14.8			
	7	28.5	17.0	–	–			
	2	27.6	15.6	–	–			
8	7	32.6	16.9	35.8	15.2			
9	1	28.1	16.5	28.9	15.5			
MEAN		29.4[d]		29.9[d']			27.2	
SEM		±0.4		±1.0			±0.5	

a = months since last phlebotomy
b = P_{50}, mmHg, prephlebotomy
b' = P_{50}, mmHg, 24 hr after phlebotomy
c = Hb in gm/dl prephlebotomy
c' = Hb in gm/dl 24 hr after phlebotomy
d = significantly greater then normal (p<0.005)
d' = significantly greater than normal (p<0.005)

Reticulocytes were not elevated in either the PV patients or the normal controls while this study was underway; however, they were not measured at a later time when the effects of an elevated Ep would be apparent. Also, the phlebotomy volume in this study was the same (480-500cc) for all the PV patients and normals, and there was no apparent effect of different blood volumes on the Ep level after phlebotomy. Miller, et al[12,13] have noted that reticulocytosis may occur after blood loss without precedent significant increases in Ep. They proposed that changes in Ep may have occurred but were not detectable by RIA because of small sample size[13].

In the present group of patients and normals, serum Ep was significantly lower in PV than in normals. However, the Hb level was higher, so these data did not support or refute the suggestion that Ep levels in PV are lower than in normal individuals at a given level of Hb[6,8]. However, the Hb of the patients with PV, as a group, had a lower affinity for oxygen than did the Hb of the normal subjects. Cassels and Morse[14] reported that blood from patients with PV had normal "in vitro" oxygen dissociation curves, which suggest that the low "in vivo" affinity for oxygen (increased P_{50}) we observed may have been due to factors extrinsic to the red cells. However, our patients also tended to show increased levels of 2,3-DPG which would implicate a red cell abnormality.

Even small changes in the position of the sigmoid curve relating oxygen tension to the saturation of hemoglobin with oxygen would affect the amount of oxygen released to tissues without affecting oxygen uptake[15].

Table 3. Calculated in vivo P$_{50}$, 2,3-DPG values and red cell indices in polycythemia vera patients and normal controls obtained on year later.

PATIENT #	MONTHS[a]	HB[b]	HCT	MCV	MCHC	RETICS[c]	2,3DPG[d]	P50[e]
1 M	2.0	16.3	49.0	85	33	2.6	17.9	30.9
2 M	0.25	18.1	55.5	69	33	2.3	9.0	27.8
	1.0	16.6	50.9	67	33	2.1	17.4	28.7
	2.0	17.0	53.7	67	32	1.2	16.5	27.5
3 M	2.0	17.0	53.8	65	32	1.6	21.6	30.4
	0.25	16.0	51.1	67	31	2.0	21.4	30.1
	1.0	15.4	49.5	64	31	4.2	22.4	31.6
4 M	6.5	15.4	50.9	73	30	0.5	16.6	29.9
5 M	4.0	15.8	51.6	65	31	1.9	29.4	29.6
6 F	4.5	16.5	53.5	94	31	3.1	21.1	27.2
	1.0	15.4	48.0	91	32	–	13.0	28.3
10 F	8.0	16.4	49.0	73	33	0.7	9.3	28.2
	8.0	16.5	49.6	74	33	0.7	13.3	28.1
	1 day	15.1	45.5	74	33	0.8	14.0	28.9
NORMALS #								
1 M		16.2	45.1	87	36	0.6	11.9	28.1
2 F		13.8	40.6	91	34	3.8	13.6	28.2
3 F		14.9	42.1	84	35	2.5	13.6	24.7
4 M	1 day	14.4	40.0	85	36	2.8	11.9	28.2
5 M		16.8	48.1	81	35	0.8	11.1	27.4

a = months since last phlebotomy
b = Hb in gm/dl
c = reticulocytes in %
d = 2,3DPG in μm/g
  (Expected normal values: Males 12.8±2.3, Females 13.6±2.6 μm 2,3DPG/g Hb)
e = calculated in vivo p50, mmHg

Hb with a lower than normal affinity for oxygen in PV could result in better oxygen release to the tissues and, therefore, lower production of Ep than would the same amount of hemoglobin in normals. This could explain the previous findings[6,8] that patients with PV may have levels of Ep in their plasma or urine that are less than normals would have with the same Hb or Hct.

ACKNOWLEDGEMENTS

The authors are grateful to Drs. H. Cutting, L. Hollander, and H. Stauffer for permission to study their patients and to C. Allen, D. Carpenter and P. Garbutt for technical assistance.

REFERENCES

1. Erslev, E.J., J. Caro, E. Kansu, O. Miller, and E. Cobbs. 1979. Plasma erythropoietin in polycythemia. Am. J. Med. 66: 243.
2. deKlerk, G., P.C.J. Rosengarten, R.J.W.M. Vet, and R. Goudsmit. 1981. Serum erythropoietin (ESF) titers in polycythemia. Blood 58: 117.
3. deKlerk, G., P.C.J. Rosengarten, R.J.W.M. Vet, and R. Goudsmit. 1981. Serum erythropoietin (ESF) titers in anemia. Blood 58: 1171.

4. Napier, J.A.F., and A. Janowska-Wieczorek. 1981. Erythropoietin measurements in the differential diagnosis of polycythemia. Brit. J. Haematol. 48: 393.

5. Cotes, P.M. 1982. Immunoreactive erythropoietin in serum. I. Evidence for the validity of the assay method and the physiological relevance of estimates. Brit. J. Haematol. 50: 427.

6. Garcia, J.F., S.N. Ebbe, L. Hollander, H.O. Cutting, M.E. Miller, and E.P. Cronkite. 1982. Radioimmunoassay of erythropoietin: circulating levels in normal and polycythemic human beings. J. Lab. Clin. Med. 99: 624.

7. Sherwood, J.B., E. Goldwasser, R. Chilcote, L.D. Carmichael, and R.L. Nagel. 1986. Sickle cell anemia patients have low erythropoietin levels for their degree of anemia. Blood 67: 46.

8. Adamson, J.W. 1968. The erythropoietin/hematocrit relationship in normal and polycythemic man: implications of marrow regulation. Blood 32: 597.

9. Berlin, N.I. 1975. Diagnosis and classification of the polycythemias. Semin Hematol. 12: 339.

10. Rorth, M. 1972. Methods for determination of the oxyhemoglobin dissociation curve. Ser. Haematol. 5(1): 48.

11. Bennington, J.L. (ed). 1984. Saunders Dictionary and Encyclopedia of Laboratory Medicine and Technology. Philadelphia, W.B. Saunders. p. 1128.

12. Miller, M.E., and D. Howard. 1979. Modulation of erythropoietin concentrations by manipulation of hypercarbia. Blood Cells 5: 389.

13. Miller, M.E., E.P. Cronkite, and J.F. Garcia. 1982. Plasma levels of immunoreactive erythropoietin after acute blood loss in man. Brit. J. Haematol. 52: 545.

14. Cassels, D.E., and M. Morse. 1953. The arterial blood gases, the oxygen dissociation curve, and the acid-base balance in polycythemia vera. J. Clin. Invest. 32: 52.

15. Finch, C.A., and C. Lenfant. 1972. Oxygen transport in man. New Engl. J. Med. 286: 407.

ERYTHROPOIETIN AND THE POLYCYTHEMIA OF

HIGH-ALTITUDE DWELLERS

Nicholas Dainiak, Hilde Spielvogel, Sandra Sorba,
Leon Cudkowicz

Departments of Medicine and Laboratory Medicine,
University of Connecticut Health Center, and
Connecticut Region Red Cross, Farmington, CT, USA

Department of Internal Medicine,
University of Cincinnati, Cincinnati, OH, USA

Instituto Boliviano de Biologia de Altura,
La Paz, Bolivia

INTRODUCTION

High-altitude natives residing at similar altitudes exhibit varying degrees of erythrocytosis, depending in part upon their geographical location[1-3]. A recent report by Winslow et al[3] suggests that hematocrit and hemoglobin levels correlate with circulating erythropoietin levels in such individuals. These authors contend that individuals with elevated erythropoietin levels may be functionally anemic even though the hematocrit and hemoglobin values are elevated.

An unusual disorder characterized by profound erythrocytosis and hyperviscosity was first described by Carlos Monge Medrano in 1925[4]. This disorder, known as Monge's disease, chronic mountain sickness or "erythemic syndrome of altitude", characteristically leads to cardio-pulmonary failure and premature death[5]. Provocatively, descent to sea level results in normalization of the hematologic values, suggesting that erythropoiesis in such individuals remains subject to normal physiological control mechanisms[6]. It has been hypothesized that over-response to physiologic stimuli might be the primary cause of chronic mountain sickness[6]. We have previously suggested that a similar mechanism is responsible for erythrocytosis in young individuals wherein the drive to produce erythropoietin in response to phlebotomy is exaggerated[7,8]. In this report, we have measured erythropoietin levels in patients with Monge's disease and healthy high-altitude dwellers. Our results suggest that chronic mountain sickness may be a heterogeneous disorder that is sometimes characterized by excessive erythropoietin production in the face of profound erythrocytosis.

MATERIALS AND METHODS

Subjects

Experimental subjects in Bolivia were recruited from local healthy residents and by referral to the Instituto Bolivio Biologica de Altura (IBBA) in La Paz (3600 meters). They included 11 healthy volunteers (nine females and two males) with a mean hemoglobin of 17.1 g/dl (range of 15.2 to 20.5 g/dl) and a mean hematocrit of 49.5% (range of 46 to 58%); and eight patients with chronic mountain sickness (two females and six males) with a mean hemoglobin of 23.6 (range of 20.6 to 27.0 g/dl) and a mean hematocrit of 71.8% (range of 64 to 81%). The ages of healthy residents and patients were similar. None of the subjects visited lower altitudes or underwent phlebotomy in the several months preceeding study.

Erythropoietin Assays

Serum was separated from whole blood immediately following phlebotomy and frozen thereafter. Prior to assay, serum samples were thawed and tested in a minimum of two (and in most cases three) different erythropoietin assays. The hypertransfused, polycythemic mouse erythropoietin bioassay was used in all cases. Four days following induction of polycythemia in CF_1 female mice by hypertransfusion, 1.00 ml of serum was subcutaneously injected into each of five animals. [59]Fe was injected into the tail veins 24 hours later, and its incorporation into erythrocytes after 24 hours was determined and compared to incorporation induced by concurrently tested normal saline and erythropoietin standards, as previously described[9]. In each case, an additional erythropoietin assay was performed, as follows.

An in vitro bioassay in splenocytes of C57BL/6J x C3H/HEJ mice which were previously injected with phenylhydrazine HCl was employed, according to methods previously described[10]. In addition, the erythropoietin radioimmunoassay of Sherwood and Goldwasser[11] was performed at the University of Chicago in a double-blind fashion. Human plasma obtained from hematologically normal residents at sea level contains less than 50 mu/ml, less than 25 mu/ml and less than 10 mu/ml for the in vivo bioassay, in vitro bioassay and radioimmunoassay, respectively.

Statistical Methods

Values were compared among groups by calculation of the geometric means and analysis of variance followed by the Newman-Keuls multible comparison test[12]. Correlation coefficients (r) were determined for results obtained in different erythropoietin assays, using standard methods. Differences were considered significant when $p < 0.05$.

RESULTS

Circulating erythropoietin levels ranged from 10 to 2,000 mu/ml in all assays for all subjects tested. In general, levels obtained in the three erythropoietin assays were comparable, although the best correlation of values was between those obtained by the in vivo bioassay in hypertransfused mice and those obtained by radioimmunoassay (see Figure 1). Values obtained with the in vivo bioassay were higher than those obtained with the radioimmunoassay, and in no case was a radioimmunoassay value higher than the corresponding in vivo bioassay value for a given sample.

Table 1 summarizes the erythropoietin levels obtained in healthy individuals and patients with chronic mountain sickness. Whereas values were higher relative to hematocrit in healthy La Paz residents compared to those obtained in residents at sea level in the United States, there was no apparent correlation between hemoglobin or hematocrit level and erythropoietin level. Erythropoietin values in patients with Monge's disease could be easily separated into two distinct groups: group I with

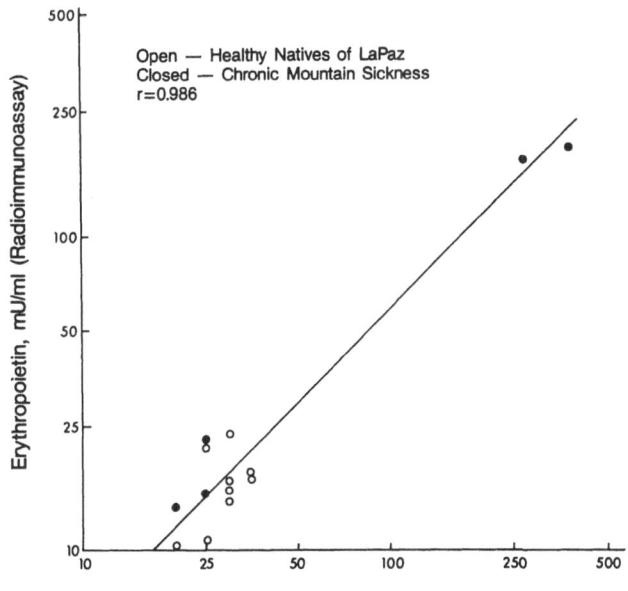

Fig. 1. Correlation of erythropoietin values obtained
in the hypertransfused mouse polycythemic in
vivo bioassay with those obtained by radioimmunoassay. Results are shown for all samples
on which both erythropoietin assays were performed. These included sera from nine healthy
residents of high altitude and from five patients
with chronic mountain sickness.

very high levels observed in sera from two patients, and group II with
levels similar to that of healthy high-altitude dwellers which were observed
in the remaining six patients.

As shown in Table 1, circulating erythropoietin values in two patients
with hemoglobin values of 21.5 and 27.0 g/dl and hematocrit values of
69% and 81%, respectively, were markedly elevated (compared to healthy
controls, $p < 0.05$) in all three erythropoietin assays (range of 173 to
2,000 mu/ml for all three assays). In contrast, erythropoietin values
in sera of the remaining six patients (whose hemoglobin levels ranged
from 20.6 to 26.0 g/dl and whose hematocrit values ranged from 64 to 79%;
compared to levels in group I, $p < 0.5$ for both hemoglobin and hematocrit
values) were similar ($p < 0.05$) to that of healthy controls (range of
14 to 50 mu/ml for all three assays). Circulating levels of the hormone
were distinct for each of the patient subgroups ($p < 0.05$).

DISCUSSION

In spite of the description of chronic mountain sickness (Monge's
disease) over sixty years ago, remarkably little is known concerning its
pathogenesis. The unusually high level of erythrocytosis that is manifest
in patients suffering from this disease suggests that a primary abnormality
of the erythropoietic axis may have developed. Until recently, accurate
measurements of erythropoietin levels and of erythroid progenitor cell numbers and/or function using biochemically defined tissue culture systems 13-16
have not been available. Here, we have evaluated erythropoietin levels

Table 1. Circulating Erythropoietin Levels*

Subjects	In Vivo Bioassay	In Vitro Bioassay	RIA
Healthy	27.7 ± 5.4	18.8 ± 9.4	16.6 ± 4.5
Chronic Mountain Sickness:			
Group I	317.5 ± 57.5	1150.0 ± 850.0	180.0 ± 7.0
Group II	24.2 ± 3.4	30.8 ± 5.0	17.7 ± 4.5

* Mean ± SD erythropoietin values (mU/ml) obtained in sera from 11 healthy volunteers and eight patients with Monge's disease (two patients in Group I plus six patients in Group II.)

in healthy and polycythemic high-altitude dwellers in order to determine whether the erythroid stimulatory factor portion of the axis is abnormal.

Our results indicate that two distinct subgroups of Monge's disease may exist. In the first, erythropoietin levels are elevated while in the second, erythropoietin levels are apparently "normal". Interestingly, one patient with chronic mountain sickness was described as having an elevated erythropoietin level of 45 mu/ml by radioimmunoassay in the recent report of Winslow et al[3]. It is unknown whether patients with "normal" erythropoietin values display abnormalities in erythroid progenitor cell sensitivity to erythropoietin and/or other erythropoietic growth factors such as interleukin-3, erythroid burst-promoting activity, GM-colony stimulating factor, insulin-like growth factor, platelet-derived growth factor and other cytokines that appear to operate over short distances within the local bone marrow environment[17-19]. Nevertheless, our results suggest that chronic mountain sickness is a heterogeneous disorder wherein the primary defect may occur at one or more sites within the erythropoietic system. This concept is consistent with the hypothesis that Monge's disease represents an over-response of normal physiologic mechanisms[6].

ACKNOWLEDGEMENTS

We thank Dr. E. Goldwasser for performing the radioimmunoassay and Tracy Arnone for typing the manuscript. Supported by NIH grants DK27071 and DK 31060 to Nicholas Dainiak.

REFERENCES

1. Hurtado, A. 1932. Studies at high altitude. Blood observations on the Indian natives of the Peruvian Andes. Am. J. Physiol. 100: 487.
2. Garruto, R.M., and J.S. Dutt. 1983. Lack of prominent compensatory polycythemia in traditional native Andeans living at 4,200 meters. Am. J. Physiol. 61: 355.
3. Winslow, R.M., K.W. Chapman, C.C. Gibson, M. Samaja, C.C. Monge, E. Goldwasser, M. Sherpa, F.D. Blume, and R. Santolaya. 1989.

Different hematologic responses to hypoxia in Sherpas and Quechua Indians. J. Appl. Physiol. 66: 1561.

4. Monge, M.C. 1925. Sobre un caso de enfermedad de Vaquez. In: Communicacion presentada a la Academia Nacional de Medicina, Lima. pp 1-6.

5. Ergueta, J., H. Spielvogel, and L. Cudkowicz. 1971. Cardio-respiratory studies in Chronic Mountain Sickness (Monge's syndrome). Respiration 28: 485.

6. Winslow, R.M., and C.C. Monge. 1987. Hypoxemia, Polycythemia, and Chronic Mountain Sickness, Baltimore, Johns Hopkins Univ. Press.

7. Dainiak, N., R. Hoffman, A.I. Lebowitz, L. Solomon, L. Maffei, and K. Ritchey. 1979. Erythropoietin-dependent primary pure erythrocytosis. Blood 53: 1076.

8. Kulkarni, V., K. Ritchey, D. Howard, and N. Dainiak. 1985. Heterogeneity of erythropoietin-dependent erythrocytosis: case report in a child and synopsis of primary erythrocytosis syndromes. Brit. J. Haematol. 60: 751.

9. Dainiak, N., V. Kulkarni, D. Howard, M. Kalmanti, M.C. Dewew, and R. Hoffman. 1983. Mechanisms of abnormal erythropoiesis in malignancy. Cancer 51: 1101.

10. Kobari, L., N. Dainiak, A. Najman, S. Kreczko, N.C. Gorin, G. Duhamel, andE. Frindel. 1987. Stimulatory effect of plasma from patients with acute nonlymphoblastic leukemia on early erythroid progenitors and pluripotent stem cells. Exp. Hematol. 15: 838.

11. Sherwood, J.B., and E. Goldwasser. 1979. A radioimmunoassay for erythropoietin. Blood 54: 885.

12. Nie, N.H., C.H. Hull, J.G. Jenkins, K. Steinbrenner, and D.H. Bent. 1975. Statistical package for the Social Sciences (2nd ed), New York, McGraw-Hill.

13. Iscove, N.N., L.J. Guilbert, and C. Weyman. 1980. Complete replacement of serum in primary cultures of erythropoietin-dependent red cell precursors (CFU-E) by albumin, transferin, iron saturated fatty acid, lecithin and cholestrol. Exp. Cell Res. 126: 121.

14. Dainiak, N. 1985. Role of defined and undefined serum additives to hematopoietic stem cell culture. In: E.P. Cronkite, N. Dainiak, P.R. McCaffrey, J.Palek, P.J. Quesenberry (eds), Hematopoietic Stem Cell Physiology, New York, Alan R. Liss, p 59.

15. Stewart, S., B. Zhu, and A.A. Axelrod. 1984. A "serum-free" medium for the production of erythropoietic bursts by murine bone marrow cells. Exp. Hematol. 12: 309.

16. Dainiak, N., S. Kreczko, A. Cohen, R. Pannell, and J. Lawler. 1985. Marrow cultures for erythroid bursts in a serum-substituted system. Exp. Hematol. 13: 1073.

17. Dainiak, N., L. Feldman, D. Sutter, C.M. Cohen, and A. Najman. 1988. Support of erythropoietic growth by proximal marrow cell-derived processes. In: E.D. Zanjani, M.T. Tavassoli, J.L. Ascensao (eds), Regulation of Erythropoiesis, New York, Spectrum, p 127.

18. Seiff, C.A. 1987. Hematopoietic growth factors. J. Clin. Invest. 79: 1549.

19. Dainiak, N. 1990. Classification of hormone and nonhormone growth factors. In: N. Abraham, G. Konwalinka (eds), Molecular Biology of Hematopoiesis, London, Intercept,

ENHANCED EFFECT OF INCREASED ERYTHROCYTE PRODUCTION RATE ON PLASMA

ERYTHROPOIETIN LEVELS OF MICE DURING SUBSEQUENT EXPOSURE TO HYPOBARIA

Carlos E. Bozzini, Rosa M. Alippi and Ana C. Barcelo

Department of Physiology
Faculty of Odontology
University of Buenos Aires
Republica Argentina

ABSTRACT

The present study was undertaken to determine plasma EPO levels in response to acute exposure to hypobaria in transfused-polycythemic mice previously exposed to different forms of erythropoietic stimulation.

The erythrocyte production rate (EPR) was stimulated by one of the following conditions: (1) discontinuous exposure to 456 mb during 4 weeks, (2) one weekly injection of phenylhydrazine (PHZ) during 4 weeks for induction of a compensated hemolytic state, and 3) three weekly injections of rHuEPO at a dose level of 1500 U/Kg body weight for 4 weeks. Treated and control mice received different volumes of packed red cells in order to obtain hematocrit values around 65% in all groups. Mice from each group were exposed to 456 mb during 14 h, 4 days after transfusion. Mice were bled immediately through cardiac puncture after removal from the hypobaric chamber. Plasma EPO titer was estimated by radioimmunoassay.

Two observations were worth noting: (1) plasma iEPO level was very low in mice with transfusion polycythemia exposed to hypobaria, the value of 12 ± 1.3 mU/ml being significantly different from that of 180 ± 18 mU/ml found in normocythemic mice similarly exposed, and (2) plasma iEPO levels during EPRs were previously stimulated by different conditions (hypobaria, PHZ and rHuEPO) were at least as high as those found in normocythemic mice similarly exposed. All these values were significantly higher than the value corresponding to polycythemic mice whose EPRs were not stimulated before exposure to hypobaria.

The results suggest that the higher EPO production seen in posthypoxic polycythemic than in transfused polycythemic mice during hypobaria would be related to either the increased EPR or the enhanced EPO production (or both) that occur during the period of hypobaric induction of polycythemia. Moreover, they suggest that hytpobaria or hypoxia per dose did not significantly contribute to the genesis of the phenomenom which, however, remains unexplained.

INTRODUCTION

We have previously reported[1] that RBC-59Fe uptake and plasma erythro-

poietin (EPO) titer during exposure to hypobaria is very much higher
in posthypoxic than in hypertransfused polycythemic mice at equals level
of polychemia. This finding, which does not fit with the current model
of the feedback regulation of EPO production, was recently confirmed
by Erslev and Caro[2].

The mechanism of this puzzling observation has not been explained
yet. Plasma EPO levels and erythrocyte production rates (EPR) were similarly
depressed in both types of polycythemic animals at the time of exposure
to hypobaria. However, posthypoxic mice, but not transfused mice, had
been previously hypoxic and, as a consequence, stimulated to increase
their EPRs in response to increased EPO production during the 3-week
period of hypobaric induction of polycythemia. Therefore, three factors
- hypobaric hypoxia, increased EPR and enhanced EPO production - could
be involved in the phenomenom under investigation.

The present study was thus undertaken to determine plasma EPO levels
in response to acute exposure to hypobaria in transfused-polycythemic
mice previously exposed to different forms of erythropoietic stimulation.

MATERIALS AND METHODS

Adult female mice of the CF1 strain weighing 22-24 g were used throughout the experiments.

The erythrocyte production rate (EPR) of experimental animals was
stimulated by one of the following conditions:

(1) Exposure to hypobaria: mice were placed into a steel chamber evacuated
by a vacuum pump and maintained at a pressure of 456 mb during 16=18
h/day for 4 weeks;

(2) Phenylhidrazine (PHZ) injections: mice received an initial dose
of 4.0 mg/100 g body wt of PHZ on day 0 of the first week, followed by
doses of 2.0 mg/100 g body wt injected at weekly intervals every 7 days
for 4 weeks for the induction of a compensated hemolytic state[3];

(3) Erythropoietin injections: mice were injected intraperitoneally
three times a week, for a total of 12 injections, with 0.1 ml of a solution,
PBS pH 7.4, containing purified recombinant human EPO (rHuEPO) (Elanex
Pharmaceuticals, Inc., Bothell, Washingtin, USA, Lot # K3F/K4D) at a
dose level of 1500 U/Kg.

Both control, untreated mice and mice with induced hemolytic state
were made polycythemic by injecting them with washed packed red cells
on two consecutive days. The hematocrit value was measured in all mice
of the remaining groups at the end of the period of stimulation of ERP.
When necessary, mice were transfused with variable volume of packed red
cells in order to keep the hematocrit value around 0.65. It was so done
because the erythropoietic response of transfused-polycythemic mice to
acute exposure to hypobaria is negatively related to the degree of polycythemia attained by the animals in response to the volume of red cells
transfused[1].

Mice from each group were exposed to hypobaria (456 mb) 4 days after
the end of the period of EPR stimulation or after the second transfusion,
depending on the specific group. Mice were removed from the chamber
after 14 h of exposure and bled immediately through cardiac puncture.
Plasma was then separated by centrifugation and stored at -20C until
assayed.

Plasma EPO titer was estimated by radioimmunoassay. Plasma samples were incubated in duplicate at 4 C for 24 h with rabbit antiserum raised against a crude EPO preparation in a total volume of 0.4 ml. Reaction diluent was phosphate-buffered saline, pH 7.4, containing 0.5% bovine serum albumin. After the 24 h incubation, 125I-EPO was separated from free 125I-EPO by the addition of goat anti-rabbit IgG. All samples were examined as a batch in a single assay to exclude interassay variability. The sensitivity of this assay is 4 mU/ml.

Statistical comparison between groups was made by the Student s t test. One-way analysis of variance (ANOVA) was used and a null hypothesis was rejected at the 0.05 level. When a statistically significant ratio was encountered, the statistical significance of differences between values were analysed by the Newman-Keuls multiple range test.

RESULTS

Mice exposed to hypobaria or injected with EPO effectively increased their EPR in response to treatment as judged by the hematocrit value which demonstrated the development of polycythemia. The evidence for the establishment of the compensated hemolytic state in PHZ-treated mice was obtained by hematocrit and reticulocyte count determinations. The hematocrit achieved its lowest value on day 2 after the initial injection of PHZ. It then started to increase and return to its pretreatment level by the 2nd week of treatment. The reticulocyte count of the PHZ-treated mice was elevated through the 4 week-injection period. Therefore, the EPR was also increased in this group.

Figure 1 shows plasma iEPO levels expressed in terms of mU/ml in normacythemic (C) and polycythemic (HT) mice exposed to hypobaria (C*) and in polycythemic mice whose EPRs were either not previously stimulated (HT*) or stimulated by chronic discontinuous exposure to hypobaria (PH*), phenylhidrazine injections (PHZ*) or rHuEPO administration (EPO*). Two observations are worth noting (a) plasma iEPO level was very low in mice with transfusion polycythemia exposed to hypobaria (HT*), the value of 12 ± 1.3 mU/ml being significantly different from that of 180 ± 18 mU/ml found in normocythemic mice similarly exposed (C*), and (b) plasma iEPO levels during exposure to hypobaria in mice with transfusion poly-cythemia whose EPRs were previously stimulated by different conditions (hypobaria, PHZ and rHuEPO) were at least as high as those found in normo-cythemic mice similarly exposed. All these values were significantly higher than the value corresponding to polycythemic mice whose EPRs were not stimulated before exposure to hypobaria.

DISCUSSION

Stimulation of erythropoiesis in response to EPO secretion occurs in posthypoxic polycythemic mice when they are re-exposed to hypobaria[1,2]. This EPO response has three puzzling characteristics: (a) it is as high as that of normal, normocythemic mice, (b) it is many times higher than that seen in transfused polycythemic mice in spite of the fact that both types of animals are equally polycythemic, and (c) it is not related to the degree of polycythemia.

Increased red cell production is also observed in hypertransfused polycythemic mice when they are acutely exposed to hypobaria. However, the response differs from that of posthypoxic polycythemic mice in that it is negatively related to the degree of polycythemia attained by the animals in response to the volume of red cells transfused[1] When the

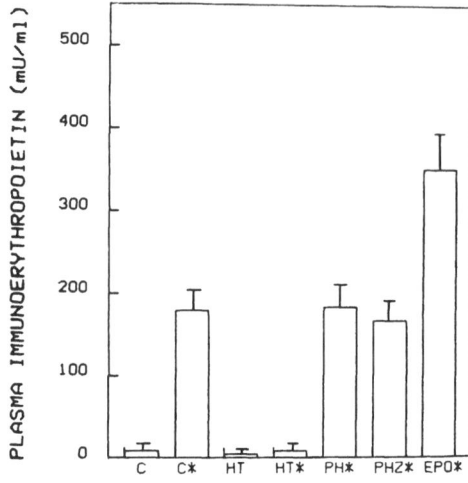

Fig. 1. Plasma immunoreactive erythropoietin concentration in normocythemic (C) and polycythemic
(HT) not exposed to hypobaria,
in normocythemic mice exposed to
hypobaria (C*) and in polycythemic
mice whose erythrocyte production
rates were either not previously
stimulated (HT*) or stimulated by
chronic discontinuous exposure to
hypobaria (PH*), phenylhydrazine
injections (PHZ*) or rHuEPO administration (EPO*). Values are Mean
± S.E.M. of 10 animals.

hematocrit value is higher than 0.65, RBC-59Fe uptake is very low and
plasma iEPO level shows a very minor increase over the hypertransfused
polycythemic control value (Figure 1).

When mice are exposed to hypobaria for as little as 30 h before
transfusion and re-exposed 4 days later, their erythropoietic responses
are as high as that of posthypoxic polycythemic mice[4]. Therefore, this
short time of exposure to hypobaria switches the erythropoietic behavior
from that of the hypertransfused polycythemic mouse to that of the posthypoxic polycythemic mouse.

From these findings, it appears that the mechanism involved in the
control of EPO synthesis and/or secretion is somewhat conditioned by
exposure to hypobaria which makes it unable to detect the polycythemic
state when the animal is re-exposed to hypobaria. As a consequence,
the control of EPO production in a posthypoxic polycythemic mouse does
not fit with the concept that EPO production is controlled solely by
the oxygen supply to the tissues relative to their needs[5].

The results of the present study, which was undertaken to dissociate
hypobaric hypoxia from EPR and plasma EPO level, indicate that the production of EPO during hypobaria is almost the same for normal mice and polycythemic mice where EPRs were stimulated before transfusion by exposure
to hypobaria, PHZ injections or rHuEPO administration and significantly
higher than in mice with transfusion polycythemia not previously erythro-
poietically stimulated. Both exposure to hypobaria and PHZ injections

increase the EPR through the production of endogenous EPO whereas EPO injections increase the EPR directly. Therefore, the results suggest that the higher EPO production seen in posthypoxic polycythemic than in transfused polycythemic mice during hypobaria would be related to either the increased EPR or the enhanced EPO production (or both) that occur during the period of hypobaria or hypoxia per dose does not significantly contribute to the genesis of the phenomenom which, however, remains unexplained.

REFERENCES

1. Alippi, R.M., A.C. Barcelo, and C.E. Bozzini. 1983. Erythropoietic response to hypoxia in mice with polycythemic induced by hypoxia or transfusion. Exp. Hematol. 11: 122.
2. Erslev, A.J., and J. Caro. 1987. Erythropoietin titers in response to anemia or hypoxia. Blood Cells 8: 1.
3. Dornfest, B.S., B.A. Naughton, R. Johnson, and A.S. Gordon. 1983. Hepatic production of erythropoietin in phenylhydrazine-induced compensated hemolytic state in the rat. J. Lab. Clin. Med. 102: 274.
4. Alippi, R.M., A.C. Barcello, and C.E. Bozzini. 1983. Enhanced erythropoiesis induced by hypoxia in hypertransfused, posthypoxic mice. Exp. Hematol. 11: 878.
5. Fried, W., L.F. Plzak, L.O. Jacobson, and E. Goldwasser. 1957. Studies on Erythropoiesis. III. Factors controlling erythropoietin production. Proc. Soc. Exp. Biol. Med. 94: 237.

METABOLISM OF RECOMBINANT HUMAN ERYTHROPOIETIN IN THE RAT

Jerry L. Spivak

Division of Hematology
Department of Medicine
The Johns Hopkins University School of Medicine
Baltimore, M.D., U.S.A.

INTRODUCTION

   Erythropoietin is the only hematopoietic growth factor which behaves
like a hormone. Produced primarily in the kidneys and to a small extent
in the liver, erythropoietin interacts with primitive erythroid progenitor
cells in the bone marrow. Under normal circumstances, the plasma concentra-
centration of erythropoietin is maintained within narrow limits but is
subject to change according to the adequacy of tissue oxygenation. Thus,
tissue hypoxia stimulates erythropoietin production while a surfeit of
oxygen suppresses it. Erythropoietin, however, is never absent from
the plasma since it is an obligatory growth factor for erythroid progenitor
cells[1].

   Erythropoietin has a molecular weight of 30,400[2] of which approximately
39% is carbohydrate[3]. Erythropoietin is rich in sialic acid residues
which are not required for expression of biologic activity in vitro[4],
but do appear to be required for expression of biologic activity in vivo[5,6].
The precise role of the carbohydrate residues of erythropoietin and in
particular, sialic acid, in the in vivo metabolism of the hormone has
not been directly examined in vivo nor are there any data on the in vivo
metabolism of recombinant erythropoietin. Therefore, we studied the
invivo behavior of recombinant human erythropoietin using a biologically
active preparation of the hormone labeled with [125]Iodide.

Materials and Methods

   Male albino Sprague-Dawley rats weighing 200-350 gms (Harlan Industries)
and pure recombinant erythropoietin (Amgen, specific activity 125,000
U/mg protein) were used for in vivo metabolic studies. The erythropoietin
was iodinated by the lactoperoxidase technique[7] to a specific activity
of 33-100 μCi/mg protein. Iodination did not impair the biological activity
of the recombinant hormone[8] as determined by an in vitro assay for erythropoietin[9]. ^{125}I-labeled human albumin (Mallinckrodt) was used as a control
for the behavior of a human plasma protein in the rat.

   Erythropoietin was desialated after labeling with [125]Iodide by either
acid hydrolysis[5] or enzymatically using immobilized neuraminidase (Sigma,
Clostridium perfringens type VI)[8]. Oxidation of the vicinal galactose
residues of the desialated erythropoietin was accomplished by exposure

Molecular Biology of Erythropoiesis
Edited by J. L. Ascensao *et al.*
Plenum Press, New York

to sodium metaperiodate as previously described[10]. Desialation did not impair the in vitro biologic activity of recombinant erythropoietin, but oxidation of desialated, recombinant erythropoietin abolished the biologic activity of the hormone as observed previously for native erythropoietin[6]. For in vivo metabolic studies, rats were anesthetized with sodium pentobarbital (40 mg/kg) by intraperitoneal injection and kept on a heating pad to maintain their body temperature between 37° and 38°C. The labeled proteins were delivered by bolus injection into the femoral vein and at selected time intervals, 0.2 ml aliquots of whole blood were collected from the tail veins in heparin. Radioactivity was measured in a gamma counter and the plasma volume of each rat was calculated from its body weight using a value of 3.93 ml/100 gm body weight for rat plasma volume[11].

For tissue accumulation studies, rats were sacrificed at selected time intervals after injection of labeled recombinant erythropoietin and the liver, spleen, kidneys and marrow from both femurs were collected. The tissues were weighed, homogenized in phosphate-buffered saline and aliquots were counted in a gamma counter. In all such studies, values were obtained for both total radioactivity and radioactivity precipitated by ice cold 20% TCA

RESULTS

The recovery of acid-precipitable radioactivity in the plasma two minutes after injection of ^{125}I-labeled recombinant human erythropoietin averaged 88% (range 66-100%) indicating that the initial volume of distribution of the hormone corresponded to the intravascular space. In this regard, its behavior was similar to that of ^{125}I-labeled human albumin. However, the subsequent plasma clearance kinetics for the two proteins were different. As shown in Figure 1A, the plasma clearance of labeled albumin conformed to a single exponential with a half-disappearance time of 210 minutes and there was no appreciable catabolism of the protein during the period of observation as measured by changes in acid-soluble plasma radioactivity in the plasma.

By contrast (Figure 1B), the plasma clearance of labeled, recombinant human erythropoietin was complex and best defined by a two compartment model[8]. As determined by the method of residuals, there was an initial distribution phase with a mean half-disappearance time of 53.1 minutes (range 30.2-81.5 minutes) and a slower elimination phase with a mean half-disappearance time of 180.1 minutes (range 124-258 minutes). There was little catabolism of the labeled hormone during the period of observation as over 90% of the plasma radioactivity remained acid-precipitable.

In Table 1, the plasma clearance of iodinated recombinant human erythropoietin in the rat is compared with previously published data for iodinated human urine erythropoietin and unlabeled rat plasma erythropoietin. Although the plasma clearance kinetics of the three types of erythropoietin are similar, the actual half-disappearance rates differ. The longer plasma survival of the recombinant or human urine erythropoietin as compared to rat erythropoietin may reflect species differences as well as the influence of iodination. The differences are not, however, attributable to variations in the biologic activity of the two human hormone preparations since the recombinant erythropoietin was biologically active while the human urine erythropoietin was not[12].

The plasma clearance of chemically-desialated erythropoietin is shown in Figure 2A and that of enzymatically desialated erythropoietin in Figure 2B. The plasma clearances were combplex and, as in the case

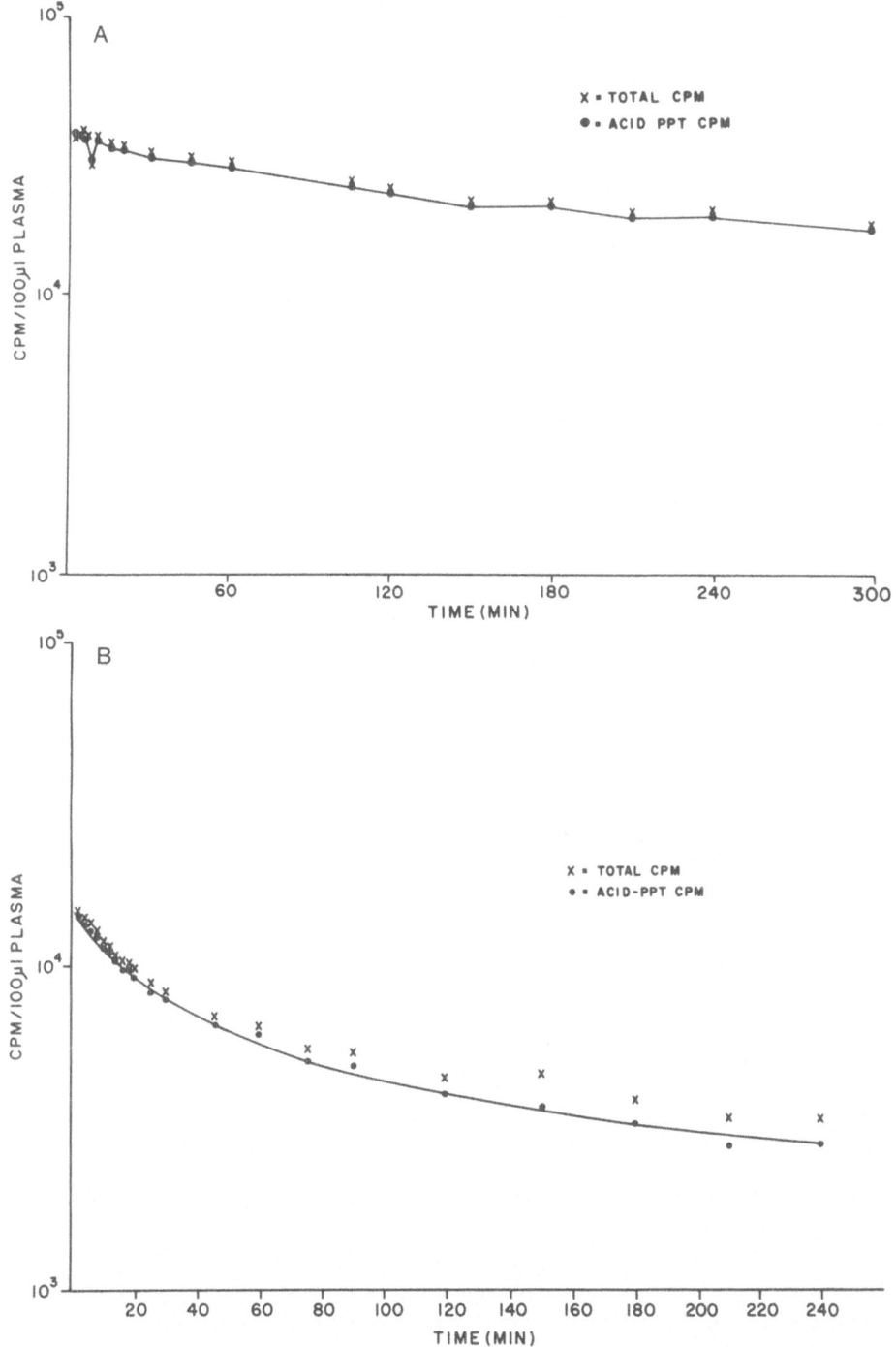

Fig. 1A. Plasma Clearance Curve for ^{125}I-labeled human albumin

Fig. 1B. Plasma clearance curve for ^{125}I-labeled human recombinant erythropoietin

of intact erythropoietin, conformed to a two compartment model. However, not only did the clearance kinetics differ but the recovery and metabolism of the desialated hormone also differed markedly from that of intact

Table 1. PLASMA CLEARANCE OF RECOMBINANT, HUMAN URINE OR HOMOLOGOUS
RAT ERYTHROPOIETIN IN THE RAT

HORMONE PREPARATION	% RECOVERY	DISTRIBUTION PHASE	ELIMINATION PHASE
Recombinant Human	88.0 ± 7.5*	53.1 ± 5.6	180.1 ± 16.3
	(66.3-100)	(30.2-81.5)	(124-258)
Human Urine (12)	--	54	204
Rat Plasma (13)	100	3.6 ± 0.5	86.0 ± 16.0

* Mean ± S.E.M.

The plasma clearance of pure recombinant human erythropoietin was measured using an
[125]I-labeled preparation of the hormone as described in Methods. The values for
pure human urine erythropoietin were also derived using an ^{125}I-labeled hormone
preparation as described in reference 12. The plasma clearance of homologous rat
plasma erythropoietin was determined by radioimmunoassay (13).

erythropoietin. As shown in Table 2, only 24% of the injected desialated
hormone was recovered from the plasma two minutes after injection and
the initial phase of plasma clearance was 25-fold faster than that of
fully sialated erythropoietin.

As illustrated in Figures 2A and 2B, the metabolism of desialated
erythropoietin was also different from intact erythropoietin, since 30
minutes after injection acid-soluble radioactivity accounted for as much
as 57% of total plasma radioactivity.

The altered plasma clearance kinetics of desialated erythropoietin
appeared to be related to exposure of the penultimate galactose residues
since oxidation of these residues by sodium metaperiodate restored both
the recovery and plasma clearance kinetics to normal (Table 2). It is
of interest in this regard that the oxidation reaction was associated
with loss of biological activity of the erythropoietin preparation in
vitro[8].

To examine the role of the liver in the plasma clearance of erythropoietin, several experiments were performed. First, rats were injected
with either desialated recombinant erythropoietin or intact recombinant
erythropoietin together with either the sialoglycoprotein, orosomucoid,
or its desialated derivative. As shown in Table 3, neither orosomucoid
or asialo-orosomucoid altered the initial plasma clearance kinetics of
intact erythropoietin. While orosomucoid also did not alter the plasma
clearance kinetics of desialated erythropoietin, desialated orosomucoid
increased the initial recovery of desialated erythropoietin and retarded
its clearance from the plasma.

To examine the interaction of recombinant erythropoietin and hepatocytes
directly, we incubated intact or desialated hormone with isolated rat
hepatocytes[14] and determined the residual hormone activity by in vitro
assay. As shown in Table 4, incubation of desialated recombinant erythropoietin, but not intact recombinant erythropoietin, with rat hepatocytes,
resulted in a decrease in residual hormone activity suggesting that the
desialated hormone had been taken up by the hepatocytes.

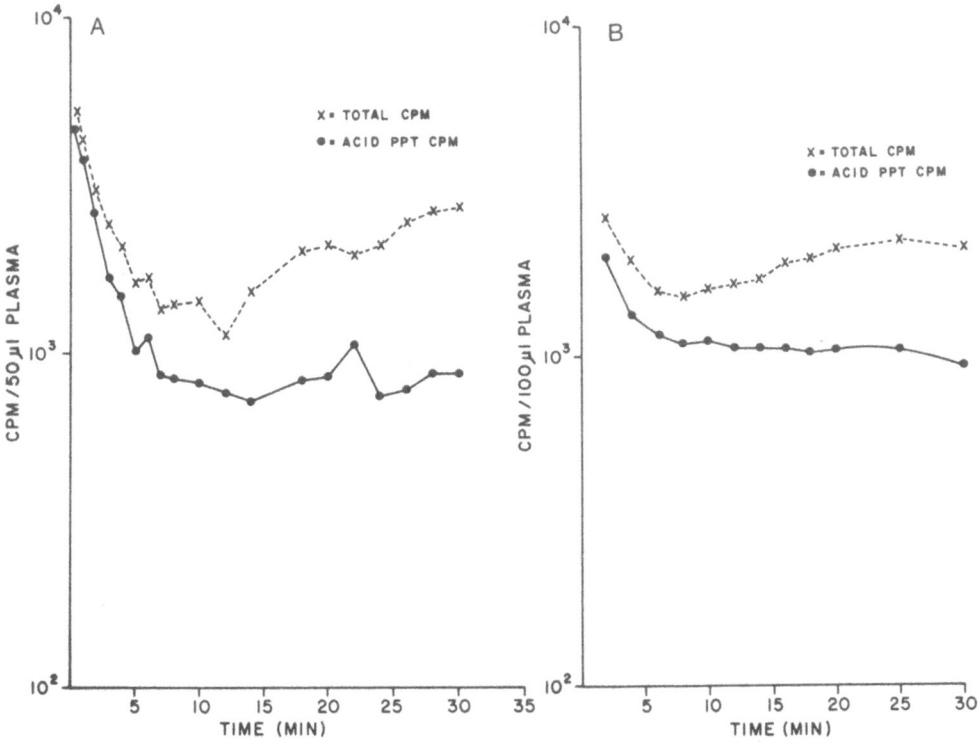

Fig. 2A. Plasma clearance curve for ^{125}I-labeled chemically desialated human recombinant erythropoietin.

Fig. 2B. Plasma clearance curve for ^{125}I-labeled enzymatically desialated human recombinant erythropoietin.

To determine the organ distribution of labeled intact and desialated erythropoietin, rats were injected with these proteins and sacrificed at selected time intervals for examination of acid-precipitable radioactivity in the liver, spleen, kidney, bone marrow (both femurs) and plasma. Labeled human albumin served as a control for organ contamination by plasma. In Table 5, the accumulation of acid-precipitable radioactivity in the individual organs is shown expressed as a fraction of the simultaneous plasma acid-precipitable radioactivity. An increase in this ratio, above 1.0, is indicative of tissue accumulation of the labeled protein[12].

As shown in Table 5, desialated erythropoietin initially accumulated rapidly in the liver and subsequently the bone marrow and spleen, but only to a small extent in the kidneys. Organ accumulation was more sluggish for intact erythropoietin and even though the organ:plasma ratios were not above 1.0, there was an evident trend in accumulation of the recombinant erythropoietin in the marrow and kidneys at 30 minutes as compared with 2 and 6 minutes. Oxidized, desialated erythropoietin failed to accumulate significantly in any tissue except the kidney (data not shown).

To examine the possibility that the plasma clearance of erythropoietin might involve phagocytosis or be mediated by receptors for sugars such as mannose or N-acetyl glucosamine, we measured the effect of administration of Dextran sulfate 500 or yeast mannan on the clearance of desialated recombinant erythropoietin. As shown in Table 6, neither substance influenced the plasma clearance of the hormone.

Table 2. EFFECT OF DESIALATION AND OXIDATION ON THE PLASMA CLEARANCE OF RECOMBINANT HUMAN ERYTHROPOIETIN IN THE RAT

HORMONE PREPARATION	% RECOVERY (ACID PRECIPI- TABLE FRACTION)	DISTRIBUTION PHASE T 1/2 (MIN)	ELIMINATION PHASE T 1/2 (MIN)
Desialated			
Chemically	23.7 ± 3.3 (6) (12.9-39.4)	2.0 ± 0.4 (0.7-3.9)	89.7 ± 22.5 (36.8-159)
Enzymatically	25.2 ± 6.0 (3) (19.2-37.3)	4.9 ± 1.3 (3.2-7.5)	90.3 ± 18.3 (55-117)
Desialated and Oxidized	100 (3)	66.7 ± 4.2 (61.7-75.1)	286.3 ± 16.6 (257.9-315.5)
Unmodified	88.0 ± 7.5 (8) (66.3-100)	53.1 ± 5.6 (30.2-81.5)	180.1 ± 16.3 (124-258)

Pure recombinant human erythropoietin was iodinated by the lactoperoxidase technique and either chemically or enzymatically desialated or enzymatically desialated and oxidized before determining its plasma clearance as described in Methods. The data represent the mean ± SEM, total number of experiments and the range of values obtained.

Table 3. EFFECT OF UNMODIFIED OR DESIALATED OROSOMUCOID ON THE INITIAL PLASMA CLEARANCE OF UNMODIFIED OR DESIALATED RECOMBINANT HUMAN ERYTHROPOIETIN IN THE RAT

COMPETITIVE INHIBITOR	ERYTHROPOIETIN PREPARATION	
	UNMODIFIED	DESIALATED
	T 1/2 (MIN)	
Orosomucoid	29.4	2.0
Desialated Orosomucoid	31.3	7.5*

* Recovery of desialated erythropoietin was 74% under these conditions. Iodinated, unmodified or chemically desialated recombinant human erythropoietin was injected into rats together with 1 mg of either unmodified orosomucoid or chemically desialated orosomucoid. The half-disappearance time for the distribution phase of plasma clearance was measured as described in Methods.

Table 4. EFFECT OF DESIALATION ON THE INTERACTION OF RECOMBINANT
ERYTHROPOIETIN AND RAT HEPATOCYTES

Treatment of Erythropoietin	^{3}H-Tdr Incorporation Into Splenic Erythroblasts (CPM/5x10^6 Cells)*	
	Before Exposure	After Exposure
None	183,000 ± 8,000 ·	---
Heated at 80 C at pH 7.4	132,000 ± 11,000	144,000 ± 12,000
Acid Hydrolysis at 80 C	171,000 ± 4,000	76,000 ± 14,000**

*   Mean ± SD
** p < .02
BKG = 7,500 ± 1,300 cpm/5x10^6 cells

One unit/ml of unmodified or desialated erythropoietin was incubated in vitro with 10^6
confluent rat hepatocytes for 24 hours at 37°C. Erythropoietin activity before and after
incubation was measured over a range of dilutions by an in vitro bioassay. Shown here are
data for 0.1 U/ml. Similar results were obtained at other dilutions.

DISCUSSION

From the data presented, a number of conclusions can be drawn. First
the initial volume of distribution of intravenously administered recombinant
human erythropoietin is the plasma volume. Second, the plasma clearance
of erythropoietin is complex and can best be described by a two compartment
model with an initial distribution phase during which the hormone equili-
brates between the intravascular and extravascular spaces and a subsequent
elimination phase. In this regard, both recombinant and human urine
erythropoietin behave identically[8,12]. The plasma clearance of homologous
rat plasma erythropoietin is similar but the half-disappearance time
of the distribution phase[13] differs from human urine[12] or recombinant
erythropoietin. Third, the sialic acid residues of erythropoietin serve
to prevent its premature removal from the plasma by the liver through
hepatocyte galactosyl receptors. Fourth, the sialic acid residues also
appear to retard the accumulation of erythropoietin in organs such as
the kidney, bone marrow and spleen. Fifth, the plasma clearance kinetics
of erythropoietin cannot wholly be explained by either its sialic acid
or its galactose residues such asialoagalactoerythropoietin has a plasma
clearance similar to intact erythropoietin. This suggests that other
determinants on the glycoprotein are involved in tissue recognition. There
was no evidence that the rapid plasma clearance of desialated erythropoietin
was due to phagocytosis or mediated through cell surface receptors for
either mannose or N acetyl glucosamine. No studies have been performed
to date concerning a role for fucose in cell recognition or erythropoietin.

The rapid uptake of desialated erythropoietin by hepatocytes can
be explained by the spatial configuration of the carbohydrate side chains
of the hormone. Hepatocyte galactosyl receptors have the highest affinity
for tetraantennary oligosaccharides containing terminal galactose residues
and in particular those which contain a 1-4 N acetyl glucosamine-mannose
linkage[15,16]. Erythropoietin is rich in N-linked tetraantennary oligosaccharides with this side-chain configuration[3]. What is unexplained, however,

Table 5. COMPARISON OF THE TISSUE UPTAKE OF UNMODIFIED AND DESIALATED RECOMBINANT HUMAN ERYTHROPOIETIN IN THE RAT

ORGAN	2 MIN		6 MIN		30 MIN	
	ASIALO-ERYTHROPOIETIN	UNMODIFIED ERYTHROPOIETIN	ASIALO-ERYTHROPOIETIN	UNMODIFIED ERYTHROPOIETIN	ASIALOERYTH-ERYTHROPOIETIN	UNMODIFIED ERYTHROPOIETIN
	$(CPM.GM \ TISSUE^{-1}/CPM.ML \ PLASMA^{-1})$					
Liver	6.0	0.10	24.0	0.1	4.0	0.2
Spleen	0.5	0.10	1.7	0.1	1.6	0.1
Kidney	0.8	0.10	3.5	0.1	1.2	0.3
Bone Marrow	0.3	0.10	1.0	0.1	1.1	0.4

Pure recombinant human erythropoietin was iodinated or iodinated and desialated before injecting into rats. At 2,6, and 30 min, the rats were sacrificed and the organs removed for measurements of acid-precipitable radioactivity as described in Methods.

Table 6. EFFECT OF YEAST MANNAN OR DEXTRAN SULFATE 500
ON THE PLASMA CLEARANCE OF DESIALATED RE-
COMBINANT HUMAN ERYTHROPOIETIN IN THE RAT

COMPETITIVE INHIBITOR	DISTRIBUTION PHASE T 12 (MIN)	ELIMINATION PHASE T 1/2 (MIN)
None	3.9	122
Dextran sulfate 500	3.3	108
Yeast Mannan	4.8	113

Rats were injected with labeled desialated erythropoietin alone or with the
erythropoietin preparation and yeast mannan. Dextran sulfate 500 was injected 2
hours before injecting the desialated erythropoietin. Plasma clearances were
determined as described in Methods.

is the difference in plasma clearance rates of intact human erythropoietin
and homologous rat plasma erythropoietin. Indeed, the plasma clearance
kinetics of rat plasma erythropoietin are similar to those of desialated
erythropoietin.

It is well established that the kidneys do not contain significant
stores of erythropoietin but rather respond to demands for the hormone
by de novo synthesis[17,18,19]. If erythropoietin synthesized on demand
is incompletely glycosylated, it would be rapidly cleared from the plasma
not only by hepatic galactosyl receptors but possibly by marrow endothelial
cells which also have galactosyl receptors[20,21]. Since the homologous
rat erythropoietin was obtained by subjecting the donor rats to hypoxia,
it is possible that erythropoietin produced in the steady-state would
behave differently. Certainly, complete glycosylation is not required
for secretion of newly synthesized erythropoitin[22] and selective differences
in hepatocyte and marrow endothelial cell galactosyl receptors could
result in differential uptake of the hormone by the two organs. The
possibility that partially desialated erythropoietin is the most active
form of the hormone has both physiologic and therapeutic implications.

REFERENCES

1.   Spivak, J.L. 1986. The mechanism of action of erythropoietin.
        Intl. J. Cell Cloning 4: 139.
2.   Davis, J.M., T. Arakawa, T.W. Strickland, and D.A. Yphantis. 1987.
        Characterization of recombinant human erythropoietin produced
        in Chinese hamster ovary cells. Biochemistry 26: 2633.
3.   Sasaki, H.B., B. Bothner, A. Dell, and M. Fukuda. 1987. Carbohydrate
        structure of erythropoietin expressed in Chinese hamster ovary
        cells by human erythropoietin cDNA. J. Biol. Chem. 262: 12059.
4.   Goldwasser, E., and CK-H. Kung. 1968. Progress in the purification
        of erythropoietin. Ann. N.Y. Acad. Sci. 149: 49.
5.   Lukowsky, W.A., and R.H. Painter. 1972. Studies on the role of
        sialic acid in the physical and biological properties of erythro-
        poietin. Canad. J. Biochem. 50: 909.
6.   Goldwasser, E., CK-H. Kung, and J. Eliason. 1974. On the mechanism
        of erythropoietin-induced differentiation. J. Biol. Chem.
        249: 4202.

7. Marchalonis, J.J. 1969. An enzymic method for the trace iodination of immunoglobulins and other proteins. <u>Biochem. J.</u> 113: 299.

8. Spivak, J.L., and B.B. Hogans. 1989. The in vivo metabolism of recombinant human erythropoietin in the rat. <u>Blood</u> 73: 90.

9. Krystal, G. 1983. A simple microassay for erythropoietin based on ^{3}H-thymidine incorporation into spleen cells from phenylhydrazine-treated mice. <u>Exper. Hematol.</u> 20: 649.

10. Spiro, R.G. 1963. Periodate oxidation of the glycoprotein fetuin. <u>J. Biol. Chem.</u> 239: 567.

11. Regoeczi, E. 1975. Hepatic uptake of asialoglycoproteins in vivo: quantification using a dual-isotope technique. <u>J. Nucl. Biol. Med.</u> 19: 149.

12. Emmanouel, D.S., E. Goldwasser, and A.I. Katz. 1984. Metabolism of pure human erythropoietin in the rat. <u>Amer. J. Physiol.</u> 247: F168.

13. Steinberg, S.E., J.F. Garcia, G.R. Matzke, and J. Mladenovic. 1986. Erythropoietin kinetics in rats: generation and clearance. <u>Blood</u> 67: 646.

14. Seglen, P.O. 1976. Preparation of isolated rat liver cells. <u>In</u> Methods in Cell Biology, vol. 13, D.M. Prescott (ed), Academic Press, New York, p. 29.

15. Baenziger, J.U. 1985. The role of glycosylation in protein recognition. <u>Amer. J. Pathol.</u> 121: 382.

16. Lee, Y.C., R.R. Townsend, M.R. Hardy, J. Lonngren, J. Arnarp, M. Haraldsson, and H. Lonn. 1983. Binding of synthetic oligosaccharides to the hepatic gal/galNAc lectin. <u>J. Biol. Chem.</u> 258: 199.

17. Schooley, J.C., and L.J. Mahlmann. 1972. Evidence for the de novo synthesis of erythropoietin in hypoxic rats. <u>Blood</u> 40: 662.

18. Bondurant, M.C., and M.J. Koury. 1986. Anemia induces accumulation of erythropoietin mRNA in the kidney and liver. <u>Molec. Cell Biol.</u> 6: 2731.

19. Beru, N., J. McDonald, C. Lacombe, and E. Goldwasser. 1986. Expression of erythropoietin gene. <u>Molec. Cell Biol.</u> 6: 2571.

20. Katoaka, M., and M. Tavassoli. 1985. Identification of lectin-like substances recognizing galactosyl residues of glycoconjugates on the plasma membrane of marrow sinus endothelium. <u>Blood</u> 65: 1163.

21. Regoeczi, E., P.A. Chindemi, M.C.W. Hatton, and L.R. Berry. 1980. Galactose specific elimination of human asialotransferins by the bone marrow in the rabbit. <u>Arch. Biochem. Biophys.</u> 205: 76.

22. Dube, S., N. Lin, R. Manger, J.W. Fisher, and J.S. Powell. 1987. Erythropoietin (EP) requires specific addition of carbohydrate (CHO) side chains for intracellular processing and secretion. <u>Blood</u> 70: 170a.

REGULATION OF EXTRARENAL ERYTHROPOIETIN PRODUCTION

Walter Fried

Rush-Presbyterian
St. Luke's Medical Center
Chicago, IL 60612

In 1953, Erslev[1] discovered that the plasma of severely anemic rabbits containes a factor that is capable of increasing the rate of erythropoiesis of normal rabbits. Three years later, Jacobson et al[2] reported that nephrectomy prevented the appearance of this factor (erythropoietin) in the plasma of anemic rats, and hypothesized that erythropoietin (Epo) is produced in the kidneys. The results of numerous studies, subsequently performed to test this hypothesis, provided substantial support for the concept that the kidneys are the primary organ of Epo production, but didn't negate the possibility that other organs could also produce Epo. Reports of detectable levels of bioactive Epo in the plasma of anephric neonatal rats[3] and of intensly hypoxic anephric adult rats[4,5] and rabbits[6] indicated that extrarenal sites of Epo exist. The significance of extra-renally-produced Epo in humans was subsequently evidenced by the observations that erythropoiesis doesn't cease in anephric patients[7]; and that the plasma of extremely anemic, anephric, persons contain detectable amounts of Epo[8,9].

In this chapter, I will review the results of studies, undertaken in order to characterize the factors that control extrarenal erythropoietin production.

During fetal and neonatal life, Epo is predominantly produced in the livers of mammals[10,11,12]. At about the time of birth, the major site of Epo production shifts to the kidneys[13,14]. Four weeks after birth, the livers of anephric rats are not able to produce more than 5-10% as much Epo as the kidneys and may not produce any Epo in adults with intact kidneys. Consequently, no one has yet succeeded in extracting detectable amounts of biologically active Epo from the liver,[15,16] although small amounts of Epo mRNA have been extracted from hepatic tissues[17,18]. Results of several experiments suggest that Kupffer cells are the source of hepatic Epo[19,20]. However, the data supporting this proposal are not conclusive.

Nathan et al[7], in their study of erythropoiesis in anephric patients, noted that the rate of erythropoiesis rose when their patients became more hypoxic. This observation suggested that extrarenal Epo production is responsive to decreases in the O_2 supply to the liver. This concept was tested in rats[21] by determining the plasma Epo titer of nephrectomized and sham operated rats, that were 1) made anemic by phlebotomy, 2) exposed to 0.5 atmosphere, 3) exposed to 0.4 atmosphere, or 4) phlebotomized and then exposed to 0.4 atmosphere for 8 hours (phlebotomy and exposure

Molecular Biology of Erythropoiesis
Edited by J. L. Ascensao *et al.*
Plenum Press, New York

to hypobaric conditions was initiated 1 hour post-nephrectomy). Plasma Epo titers were assayed in hypertransfused mice; and the results are expressed in terms of ^{59}Fe uptake into RBC's of assay mice (see Fig. 1). This data indicates that both renal and extrarenal sites respond to increased intensity of hypoxia by producing more Epo. Accordingly, extrarenal sites of Epo production are primarily regulated by O_2 requirements: O_2 supply, but at any given setting of this ratio, they produce only a fraction of the amount of Epo produced by the kidneys.

We subsequently compared the effects of various experimental manipulations on renal and extrarenal Epo production. The results of these studies are summarized in Table 1. It is noteworthy that, unlike renal Epo production, that from extrarenal sites is not suppressed by protein deprivation, nor is it stimulated by androgenic steroids. Nevertheless, males produce more extrarenal Epo than females when exposed to comparable conditions. Consequently, the effect of sex differences on Epo production is not simply due to differences in androgen secretion.

The aforementioned studies indicate that extrarenal sites of Epo production respond to hypoxic stimuli. However, the maximum response is only a fraction of that elicited from the kidneys. There are, on the other hand, two sets of circumstances which result in a very substantial increase in the amount of Epo production by the liver in response to hypoxic stimuli. These occur in rats that are recovering from hepatic injury and in rats that are perfused with angiotensin II.

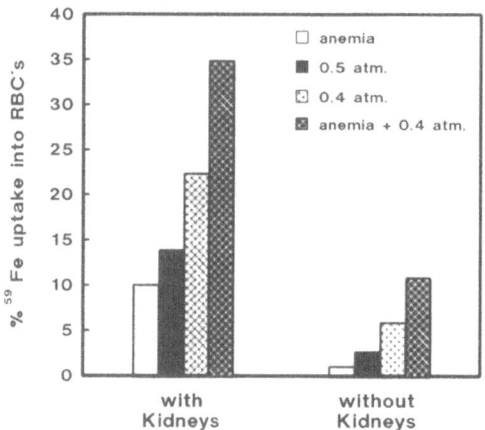

Fig. 1.   The effects of anemia and/or hypoxia on renally and extrarenally produced Epo.

The data is expressed as the mean % ^{59}Fe uptake into RBC's of 6 assay mice injected with pooled plasma from each experimental group. At the time these experiments were performed, we were not yet covering the results to milliunits by comparison to a standard curve.

This data is derived from studies, results of which were published in reference 21.

Table 1. Comparison of Effects of Various Conditions on Renal and Extrarenal EPO Production

	RENAL Epo	EXTRARENAL Epo	REFERENCE
HYPOXIA	↑	↑	21
PROTEIN DEPRIVATION	↓	— OR ↑	22, 23, 24
COBALT	↑	↑	25, 26
ANDROGENS	↑	—	26, 27, 28
SEX DIFFERENCES	Males produce more Epo than Females		14,16

The hematocrits of patients with end stage renal disease occasionally rise during the acute phase of hepatitis[29]. One explanation of this phenomenon is that the ability of the liver to produce Epo increases when it is subjected to inflammatory reactions, or while it is regenerating from hepatocellular damage[29]. To experimentally simulate reversible liver damage and determine its effect on extrarenal Epo, rats were subjected to one of the following procedures: 1) removal of 60% of the liver, 2) CCL$_4$ administration or 3) ligation of the common bile duct. At various times afterwards, batches of rats in each group were nephrectomized and exposed to 0.425 atmosphere for 6 hours. The plasma Epo titer was then assayed in post-hypoxic polycythemic mice. The results are shown in Fig. 2. In both CCL$_4$ treated and partially hepatectomized rats, the amount of Epo produced in response to hypoxia, fell immediately after the liver was damaged. It then rose to levels that substantially exceeded

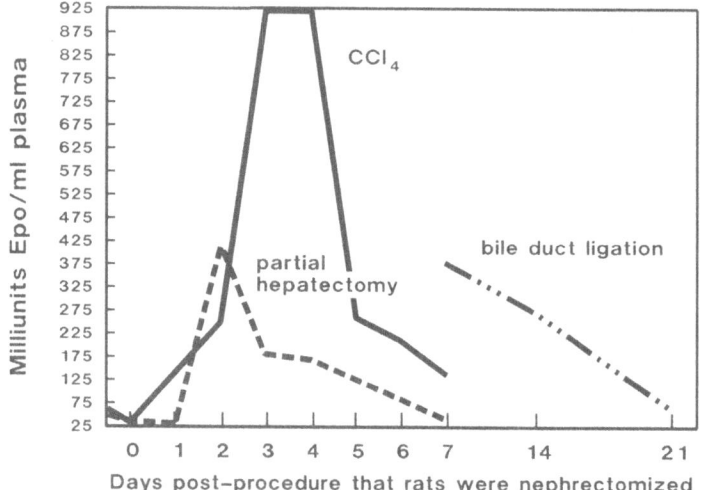

Fig. 2.   The effects of various types of liver injury on the plasma Epo titer of rats made hypoxic one hour post-nephrectomy.

The data shown in this figure were previously published, in part, in references 30, 31, 32, and 33.

that in controls, and again returned to control levels after the liver was restored to near normal. In rats with ligated bile ducts, the rise in the plasma Epo level during hypoxia exceeded that of controls throughout the study period. Neither CCL$_4$ nor partial hepatectomy caused significant changes in the plasma Epo levels of hypoxic non-nephrectomized rats. In all three models, Kupffer cell hyperplasia, hepatocyte proliferation, and even bile duct epithelial cell proliferation are demonstrable, during the time interval after the procedure, when the capacity for extrarenal Epo production is maximal. Consequently, any or all of these changes may contribute to the observed increase in hepatic Epo production. I am unaware of any further experimental attempts to identify the variable that is responsible for this phenomenon.

Infusion of angiotensin II substantially increases the amount of Epo produced by anephric rats in response to hypoxia[35]. To put this phenomenon in proper perspective, it is necessary to provide some background information that led to this observation. All of the data on extrarenal Epo, discussed to this point, have been generated in rats that are made hypoxic within an hour after removal of both kidneys. If, on the other hand, hypoxia is initiated more than 8 hours post-nephrectomy, extrarenal Epo production declines to a barely detectable level[36]. Peschle, et al[37] reported that rats, injected 18 hours post-nephrectomy with renal extracts that contain the hypothetical "renal erythropoietic factor", produce as much Epo as rats made hypoxic within an hour post-nephrectomy. Subsequently, it was observed that this phenomenon could be reproduced using renal extracts with high renin activity; or by injection of large boluses of angiotensinII[38,39]. To explain these observations, we hypothesized that nephrectomy eliminates, not only the major source of Epo production but also renin, which is required for continued generation of angiotensin II. As the tissue stores of angiotensin II decline post-nephrectomy, the liver loses its ability to produce Epo in response to hypoxia. We undertook the following series of experiments to determine how angiotensin II administration affects extrarenal Epo production in rats made hypoxic either one or 18 hours post-nephrectomy. Results are shown in Fig. 3. Miniosmopumps, loaded with either angiotensin II (amount was calculated to deliver 5 micrograms per hour for at least 7 days) or saline were implanted subcutaneously into rats. Half of the rats in each group were then nephrectomized, and the other half were nephrectomized seventeen hours later. One hour afterwards, all rats were exposed to 0.425 atmosphere for 6 hours. Their plasma Epo titers were then assayed. The plasma Epo titers of saline treated rats rose to detectable levels only in the group that was nephrectomized one hour prior to exposure to hypoxia; whereas that of angiotensin-treated rats rose substantially whether they were made hypoxic one or eighteen hours post-nephrectomy. As a matter of fact, the plasma Epo titers of angiotensin-treated rats, made hypoxic 18 hours post-nephrectomy, were several-fold higher than those of rats made hypoxic one hour post-nephrectomy. The data illustrated in Fig. 3

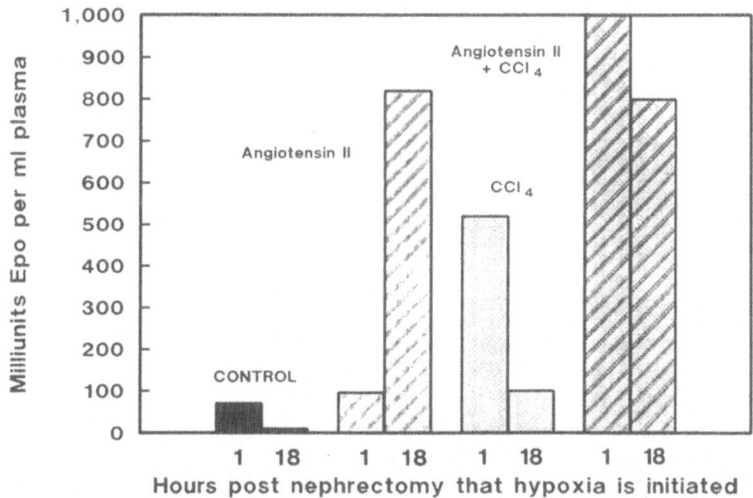

Fig. 3. The effects of Angiotensin II and CCL4, given separately or together, on the plasma Epo levels of rats made hypoxic one or 18 hours post-nephrectomy.

The data shown in this figure were originally published in reference 35.

also compares the plasma Epo titers if CCL_4 treated rats, angiotensin-treated rats, rats treated with both agents, and untreated rats that were made hypoxic one or 18 hours post-nephrectomy. Whereas angiotensin-treated rats had higher plasma Epo titers if hypoxia was initiated 18 hours post-hypoxia, those of CCL_4 treated and control rats were much lower if hypoxia was initiated 18 hours as compared to one hour post-nephrectomy. The highest plasma Epo titers of about 1000 milliunits/ml were observed in angiotensin-treated rats made hypoxic 18 hours post-nephrectomy and CCL_4 + angiotensin treated rats made hypoxic one hour post-nephrectomy. This probably reflects the maximum rate of Epo production by the adult rat liver, and is comparable to the amount of Epo produced by rats, with functioning kidneys, that are exposed to 0.5 atmosphere. (See Table 2). These studies indicate that extrarenal sites produce only about 2.5% as much Epo as do renal sites, when exposed to intense hypoxia immediately post-nephrectomy, but this can be enhanced up to ten-fold by liver injury and/or infusion of angiotensin. Unless angiotensin is infused post-nephrectomy, extrarenal sites lose their ability to produce Epo within 18 hours post-nephrectomy.

To directly test the hypothesis that extrarenal Epo production requires the presence of angiotensin II, we administered the angiotensin converting enzyme inhibitor, captopril, to rats and determined its effect on extrarenal Epo production. All rats, in the study, received CCL_4 to increase extrarenal Epo production. Half of the rats also received 35 mgm/Kgm of captopril dissolved in 5% glucose, twice daily for three days and again just prior to nephrectomy. The other half received 5% glucose according to the same schedule. One hour post-nephrectomy, all rats were exposed to 0.425 atmosphere for 7 hours. Their plasma Epo titers were then measured. Two other groups of rats were treated with CCL_4 and either captopril or 5% glucose, as above. They also had miniosmopumps, containing angiotensin II, implanted 18 hours prior to nephrectomy. The results shown in Fig. 4 indicate that captopril reduced the plasma Epo titer to about one-half of that of rats that received the diluent only (5% glucose solution). However, captopril did not significantly affect plasma Epo levels of rats that were simultaneously infused with angiotensin II. The observation, that captopril reduces extrarenal Epo production and that the effect

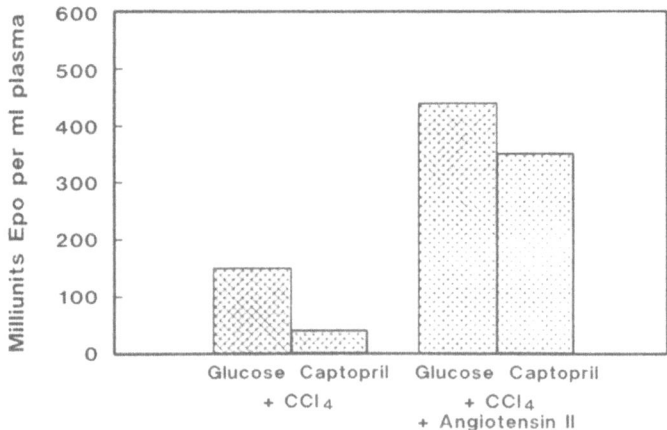

Fig. 4.   The effect of captopril on the plasma Epo titers of rats that were nephrectomized and made hypoxic three days after receiving CCL_4.

The data in this figure were originally published in part in reference 40.

Table 2.   Comparative Effects on Hypoxia on Renal and Extra-
          Renal EPO Production

	PLASMA EPO TITER HYPOXIA COMPARED TO THAT OF NORMAL NON-HYPOXIC RAT
NON-OPERATED RATS EXPOSED TO 0.425 ATMOSPHERE	**250X**
NON-OPERATED RATS EXPOSED TO 0.5 ATMOSPHERE	**50X**
NEPHRECTOMIZED RATS EXPOSED TO 0.425 ATMOSPHERE	**5X**
NEPHRECTOMIZED-CCL$_4$-ANGIOTENSIN TREATED RATS EXPOSED TO 0.425 ATMOSPHERE	**50X**

THIS TABLE SHOWS APPROXIMATE ORDERS OF MAGNITUDE THAT WE HAVE OBSERVED IN OUR LABORATORY.

is reversable by simultaneously infusing angiotensin II, is compatible
with the premise that angiotensin II is required for optimal extrarenal
Epo production. Unfortunately, tissue angiotensin II levels were not
measured. Therefore, it is not possible to ascertain whether the failure
to totally inhibit extrarenal Epo production was the result of incomplete
depletion of tissue angiotensin II stores by the dosage and schedule
of captopril administration used in the study or whether angiotensin
II is not an absolute requirement for extrarenal Epo production.

Angiotensin II, a potent vasoconstrictor, most likely affects hepatic
Epo production by altering hepatic blood flow, and thereby hepatic O_2
supply. This explanation is consistent with the observation that angiotensin
II enhances renal Epo production, only when given in doses that reduce
renal blood flow[41]. To test this hypothesis, we compared the plasma
Epo titers, mean blood pressure, and hepatic blood flow of anephric rats
that received infusions of angiotensin II, norepinepherine, or control
substances. In these studies, results of which are presented in Table 3,
norepinepherine was infused at a rate of 30 micrograms per hour and angiotensin II at a rate of 5 micrograms per hour. Mean blood pressure was
measured via tail cuff. Total hepatic blood flow (portal vein and hepatic
artery blood flow cannot be distinguished by this method) was assessed
by infusing ^{99}Tc Sulfa Colloid via tail vein, measuring the plasma clearance,
the total accumulation in the liver over 15 minutes, and calculating
the amunt extracted by the liver. The results, shown in Table 3, indicate
that angiotensin II caused a smaller increase in the systemic blood pressure
than norepinepherine. However, it caused greater reduction in hepatic
blood flow (significantly smaller percentage of ^{99}Tc Colloid was extracted
by the liver) and a markedly greater increase in extrarenal Epo production.

Although these results support the hypothesis that angiotensin II
increases extrarenal Epo production by decreasing hepatic blood flow,
alternative explanations for the phenomenon should also be considered.
Angiotensin II results in increased secretion of prostaglandins from
endothelial cells. These, in turn, have vasoactive effects and have
been shown to enhance both renal and extrarenal Epo production[42,43,44].
Therefore, it is possible that the effects of angiotensin II on extrarenal
Epo production is mediated by its effect on prostaglandin secretion in
the liver. To determine whether the erythropoietic effects of angiotensin
II are mediated by prostaglandins, the following experiments were performed
to test the effect of indomethacin, a cyclo oxygenase inhibitor, on the
erythropoietic response to angiotensin II: rats, implanted with miniosmopumps containing either angiotensin II or saline, received two intraperi-

45

Table 3. Effect of Angiotensin and of Norepinephrine on Extrarenal Ep, Mean Blood Pressure, and Hepatic Blood Flow.

	Plasma Ep* (mU/ml)	Mean Blood Pressure (measured via tail cuff)	Clearance of ^{99}Tc Colloid from Plasma (T ½ in minutes)	Amount of ^{99}Tc Colloid Extracted by Liver (as % of injected dose)
Control	71	119 (4)**	0.45 (3)	96.3 (5)
Angiotensin (5 micrograms per hour)	440	125 (5)	1.0 (4)	76.5 (7)
Norepinephrine) (30 micrograms per hour)	50	154 (6)	0.51 (4)	86.3 (7)

* All rats were nephrectomized one hour before being made hypoxic (0.43 atmosphere) for 6 hours. Their plasma was then collected and pooled for Ep assay.

** Numbers in parenthesis indicate the number of rats studied.

Data in this table is also being published in reference 33.

toneal injections (16 hours prior to and again at the time of nephrectomy) of either 10 mg/Kgm indomethacin disolved in 0.1 molar sodium carbonate, or sodium carbonate alone. One hour post-nephrectomy, they were exposed to 0.425 atmosphere for six hours. The Epo content of their plasma was then assayed and their livers were removed and assayed for PGE_2 and PGF_{2A} content. The results, shown in Table 4, indicate that rats, treated with indomethacin alone as well as those treated with angiotensin II, had substantially higher plasma Epo titers than controls. Rats, that received both indomethacin and angiotensin II, had even higher plasma Epo titers than those that received only indomethacin or angiotensin II. Consequently, the erythropoietic actions of angiotensin II are not mediated by prostaglandins. The stimulatory effect of indomethacin on extrarenal Epo production was not anticipated, since indomethacin has been reported to suppress renal Epo production in hypoxic dogs[46] and to have no effect on that of hypoxic rats with intact kidneys[45].

Naughton et al[44] concluded from their studies in anephric rats that PGE_2 increase extrarenal Epo production. However, their data also indicated that PGF_{2A} is a potent inhibitor of extrarenal Epo production. The data shown in Table 4 indicates that indomethacin causes the hepatic PGF_{2A} titer to fall to a greater extent than the hepatic PGE_2 titer. These data suggest that indomethacin increases extrarenal Epo production by decreasing the ratio of inhibiting to stimulating prostaglandins in the liver. To provide further support for this concept, we studied the effect of infusing PGF_{2A} and PGE_2 on the plasma Epo titers of anephric hypoxic rats. Miniosmopumps containing either PGF_{2A}, PGE_2, or saline were implanted subcutaneously into rats three days prior to nephrectomy and exposure to 0.425 atmosphere for 6 hours. Their plasma Epo titers were then measured. The results, shown in Figure 5, indicate that the plasma Epotiters of PGF_{2A} treated rats were significantly lower than those of controls, whereas those of PGE_2 treated rats were not significantly different from controls.

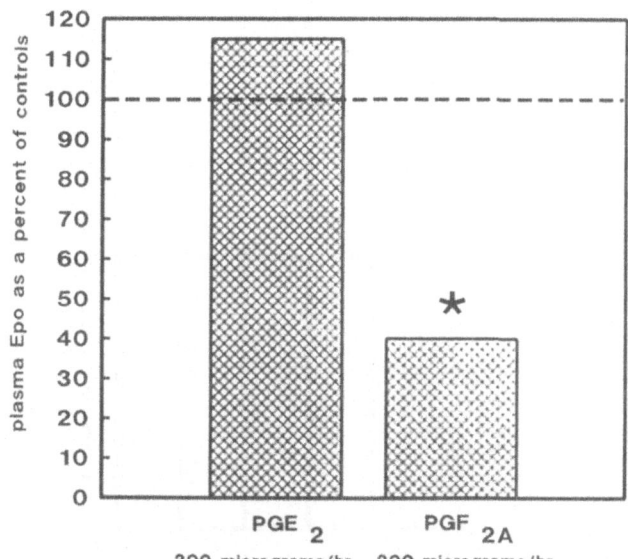

Fig. 5. The effect of prostaglandins E_2 and F_{2A} on plasma Epo levels of rats made hypoxic one hour post-nephrectomy.

The data in this figure were originally published in reference 45.

Table 4. Effects of Indomethacin and Angiotension on Extrarenal Ep and
on Hepatic Prostaglandin Levels

GROUP	PLASMA EPO LEVEL (mU/ml)	HEPATIC PGE$_2$ (ng/gm)	HEPATIC PGF$_{2A}$ (ng/gm)
CONTROL	150	250	160
INDOMETHACIN	540	190	88
ANGIOTENSIN II	300	230	160
INDOMETHACIN + ANGIOTENSIN II	1540	210	110

Data shown in this table were originally published in reference 45.

In contrast to these results, Naughton et al[44] found that PGE_2 significantly increased plasma Epo titers of anephric rats. This difference in results may be accounted for by the fact that Naughton et al[44] injected PGE_2 as a single large bolus, 24 hours prior to exposing rats to hypoxia, whereas we administered it by continuous subcutaneous perfusion.

We conclude that extrarenal sites of EPO production are located predominantly in cells within the liver that have not yet been identified. These sites are the major source of Epo during fetal life. Postnatally, the kidneys begin to take over this task and by 4 weeks of age, extrarenal sites contribute little, if any, Epo as long as the kidneys are in place and functioning. Post-nephrectomy, the adult liver can produce a small amount of Epo, that is sufficient to maintain the hematocrit at a low level in humans. The rate of extrarenal Epo production increases in response to hypoxia but is always a fraction of that produced by comparably hypoxic kidneys. Three sets of conditions have been shown to increase the capacity of extrarenal sites to produce Epo in response to hypoxia. These are: 1) hepatic injury, 2) infusion of angiotensin II and 3) injection of indomethacin.

Various types of hepatic injury can result in approximately a 5-fold increase in the plasma Epo titer after exposure of anephric rats to hypoxia. This effect seems to be temporally related to the repair process, but the etiology of the increase is extrarenal Epo production is not clear.

Infusion of angiotensin II into anephric rats can also increase the plasma Epo titer, after exposure to hypoxia, by about 5-fold. This is probably due to reduction in hepatic blood flow, and therefore decreased O_2 deliver. It is noteworthy that angiotensin II, unlike hepatic injury, increases plasma Epo titers even if hypoxia is initiated 18 hours after nephrectomy; it's presence may be required to permit optimal extrarenal Epo production to occur.

Indomethacin administration also increases the plasma Epo titer of hypoxic anephric rats several fold. It probably does so, by reducing the hepatic production of inhibitory prostaglandins, such as PGF_{2A}, relative to those that stimulate Epo production.

Since the production of pharmacologic amounts of recombinant human Epo has become a reality, there is less motivation to understand the control of extrarenal Epo for its potential therapeutic value. However, extrarenal Epo production is a physiologic phenomenon, and an understanding of how it is regulated will surely enhance our knowledge of mammalian biology.

REFERENCES

1. Erslev, A.J. 1953. Humoral regulation of red cell production. Blood 8: 349-357.
2. Jacobson, L.O., E. Goldwasser, W. Fried, and L. Plzak. 1957. Role of the kidney in erythropoiesis. Nature 179: 633-634.
3. Carmena, A.O., D. Howard, and F. Stohlman. 1968. Regulation of erythropoietin XXII. Production in the newborn animal. Blood 32: 376-382.
4. Mirand, E.A., and T.C. Prentice. 1957. Presence of plasma erythropoietin in hypoxic rats with and without kidneys and/or spleen. Proc. Soc. Exp. Biol. Med. 96: 49-57.
5. Gallagher, N.I., J.M. McCarthy, and R.D. Lange. 1961. Erythropoietin production in uremic rabbits. J. Lab. Clin. Med. 57: 281-289.
6. Erslev, A.J. 1958. Erythropoietin function in uremic rabbits. Arch. Int. Med. 101: 407-412.

7. Nathan, D.G., E. Schupak, and F. Stohlman. 1964. Erythropoiesis in anephric man. J. Clin. Invest. 43: 2158-2164.

8. Mirand, E.A., G.P. Murphy, R.A. Steeves, H.W. Weber, and F.P. Retief. 1968. Extra-renal production of erythropoietin in man. Acta. Haemat. 39: 359-365.

9. Naets, J.P., and N. Wittek. 1968. Presence of erythropoietin in the plasma of one anephric patient. Blood 31: 249-251.

10. Fried, W. 1972. The liver as a source of extrarenal erythropoietin. Blood 40: 671-677.

11. Schooley, J.C., and L.J. Mahlmann. 1974. Extrarenal erythropoietin production by the liver in the weaning rat. Proc. Soc. Exp. Biol. Med. 145: 1081-1083.

12. Zanjani, E.D., J. Poster, L.M. Burlington, L.I. Mann, and L.R. Wasserman. 1977. Liver as the primary site of erythropoietin formation in the fetus. J. Lab. Clin. Med. 89: 640-644.

13. Zanjani, E.D., J.L. Ascensao, P.B. McGlave, M. Banisadre, and R.C. Ash. 1981. Studies on the liver to kidney switch of erythropoietin production. J. Clin. Invest. 67: 1183-1188.

14. Wang, F., and W. Fried. 1972. Renal and extrarenal erythropoietin production in male and female rats of various ages. J. Lab. Clin. Med. 79: 181-186.

15. Caro, J., L. Zon, R. Silver, O. Miller, and A.J. Erslev. 1983. Erythropoietin in liver tissue extracts and in liver perfusates from hypoxic rats. Am. J. Physiol. 2224: E431-434.

16. Fried, W., J. Barone-Varelas, and T. Barone. 1982. The influence age and sex on erythropoietin titers in the plasma and tissue homogenates of hypoxic rats. Exp. Hematol 10: 472-477.

17. Bondurant, M.D., and M.J. Koury. 1986. Anemia induces accumulation of erythropoietin mRNA in the kidneys and liver. Mol. Cell Biol. 6: 2731-2733.

18. Beru, N., J. McDonald, and E. Goldwasser. 1987. Expression of the erythropoietin gene. Blood Cells 13: 263-268.

19. Paul, P., J.A. Rothmann, J.T. McMahon, and A.S. Gordon. 1984. Erythropoietin secretion by isolated rat Kupffer cells. Exp. Hemat. 12: 825-830.

20. Rich, I.N., W. Heit, and B. Kubanek. 1982. Extrarenal erythropoietin production by macrophages. Blood 60: 1007-1018.

21. Fried, W., T. Kilbridge, S. Krantz, T.P. McDonald, and R.D. Lange. 1969. Studies on extrarenal erythropoietin. J. Lab. Clin. Med. 73: 244-248.

22. Reissman, K.R. 1964. Protein metabolism and erythropoiesis, II. Erythropoietin formation and erythroid responsiveness in protein-deprived rats. Blood 23: 146-153.

23. Anagnostou, A., S. Schade, M. Ashkinaz, J. Barone, and W. Fried. 1977. Effect of protein deprivation on erythropoiesis. Blood 50: 1093-1977.

24. Anagnostou, A., S. Schade, J. Barone, and W. Fried. 1978. Effect of protein deprivation on extrarenal erythropoietin production. Blood 51: 549-553.

25. Goldwasser, E., L.O. Jacobson, W. Fried, and L. Pizak. 1958. Studies on erythropoiesis V. The effect of cobalt on the production of erythropoietin. Blood 13: 55-60.

26. Fried, W., and T. Kilbridge. 1969. Effect of testosterone and cobalt on erythropoietin production in anephric rats. J. Lab Clin. Med. 74: 623-629.

27. Fried, W., and C.W. Gurney. 1965. Erythropoietic effect of plasma from mice receiving testosterone. Nature 206: 1160-1161.

28. Mirand, E.A., A.S. Gordon, and J. Wenig. 1965. Mechanism of testosterone action on erythropoiesis. Nature 206: 270-272.

29. Brown, S., J. Caro, A.J. Erslev, and T. Murray. 1980. Spontaneous

increase in EPO and hematocrit value associated with transient liver enzyme abnormalities in an anephric patient undergoing dialysis. Am. J. Med. 68: 280-283.

30. Naughton, B.A., S.M. Kaplan, M. Roy, and A.S. Gordon. 1977. Hepatic regeneration and erythropoietin production in the rat. Science 196: 301-302.

31. Anagnostou, A., S. Schade, J. Barone, and W. Fried. 1977. Effect of partial hepatectomy on extrarenal erythropoietin production in rats. Blood 50: 457-462.

32. Fried, W., J. Barone, S. Schade, and A. Anagnostou. 1979. Effect of carbon tetrachloride on extrarenal erythropoietin production. J. Lab. Clin. Med. 93: 702-705.

33. Fried, W. (In Press). Factors that affect the rate of erythropoietin production by extrarenal sites. Annals NY Acad. of Sciences.

34. Peschle, C., G. Marone, A. Genovese, I.A. Rappaport, and M. Condorelli. 1976. Increased erythropoietin production in anephric rats with hyperplasia of the reticuloendothelial system induced by colloidal carbon or zymosan. Blood 147: 325-337.

35. Fried, W., J. Barone-Varelas, T. Barone, and A. Anagnostou. 1982. Effect of angiotensin infusion on extrarenal erythropoietin production. J. Lab. Clin. Med. 99: 520-525.

36. Schooley, J.C., and L.J. Mahlmann. 1972. Erythropoietin production in the anephric rat 1. relationship between nephrectomy and time of hypoxic exposure and erythropoietin production. Blood 33: 31-38.

37. Peschle, C., G.F. Sasso, I.A. Rappaport, and M. Condorelli. 1972. Erythropoietin production in nephrectomized rats: possible role of the renal erythropoietic factor. J. Lab. Clin. Med. 79: 950-959.

38. Gould, A.B., S.A. Goodman, and D. Green. 1973. An in-vivo effect of renin on erythropoietin formation. Lab. Invest. 28: 719-722.

39. Anagnostou, A., R. Baranowski, V.K.G. Pillay, N. Kurtzmann, G. Vercellotti, and W. Fried. 1976. Effect of renin on extrarenal erythropoietin production. J. Clin. Med. 88: 707-715.

40. Fried, W., J. Barone-Varelas, and C. Morley. 1984. Factors that regulate extrarenal erythropoietin production. Blood Cells 10: 287-304.

41. Malgor, L.A., and J.W. Fisher. 1969. Antagonism of angiotensin by hydralazine on renal blood flow and erythropoietin production. Am. J. Physiol. 216: 563-566.

42. Dukes, P.P., N.A. Shore, D. Hammond, J. Ortega, and M.C. Data. 1973. Enhancement of erythropoiesis by prostaglandins. J. Lab. Clin. Med. 82: 704-712.

43. Schooley, J.C., and L.J. Mahlmann. 1978. Stimulation of erythropoiesis in plethoric mice by prostaglandins and its inhibition by anti-erythropoietin. Proc. Soc. Exp. Biol. Med. 138: 523-524.

44. Naughton, B.A., G.K. Naughton, P. Liu, J.M. Arce, S. Pilero, and A.S. Gordon. 1982. The effects of prostaglandins on extrarenal erythropoietin production. Proc. Soc. Exp. Biol. Med. 170: 231-236.

45. Fried, W., C. Morley, J. Barone-Varelas, A. Bidani, and A. Prancan. 1988. Effect of indomethacin and of prostaglandins on extrarenal erythropoietin production. J. Lab. Clin. Med. 111: 184-188.

46. Gross, D.M., W. Jubiz, and J.W. Fisher. 1977. Released erythropoietin and prostaglandin E by the dog kidney during hypoxic hypoxia and the effect of indomethacin. Fed. Proc. 36: 1052.

STRUCTURE AND ROLE OF CARBOHYDRATE IN HUMAN ERYTHROPOIETIN

Minoru Fukuda, Hiroshi Sasaki*, and Michiko N. Fukuda

La Jolla Cancer Research Foundation
La Jolla, California 92037, U.S.A.

*Fuji-Gotemba Research Laboratories
Chugai Pharmaceuticals Co., Ltd.,
Shizuoka, Japan

Erythropoietin is a glycoprotein hormone with a molecular weight of 31 kilodaltons, with 40% of its molecular mass accounted for by carbohydrates. These carbohydrates have been shown to consist of one O-linked with three N-linked oligosaccharides[1,2]. Erythropoietin is synthesized in the kidney and circulates in the blood to stimulate red cell proliferation and differentiation in the bone marrow. This hormonal activity is abolished in vivo after sialic acid residues are removed[3,4]. But, the asialo-erythropoietin has full activity when the assay is done in vitro[5]. The apparent loss of activity of the asialo-erythropoietin in vivo can be explained by the rapid clearance from the circulation by hepatic cells[5]. When the terminal sialic acid residues are removed from the oligosaccharides, galactose residues become the new terminal sugars. These galactose-terminated glycoproteins may then be recognized by receptors present in hepatocytes and are internalized by endocytosis, followed by digestion with lysosomes[6]. Thus, it is apparent that the erythropoietin should be fully glycosylated, in particular, sialylated well to accomplish the hormone action in vivo.

Human peptide hormones can be produced abundantly by using recombinant DNA expression vectors in cultured cells. In order to acquire proper carbohydrate moiety, glycoprotein hormones must be expressed in animal cells other than bacteria or yeast[7,8]. This is because bacteria lack glycosyltransferases to attach carbohydrate chains to polypeptides, and yeast attaches only high-mannose oligosaccharides, which will be rapidly cleared from blood circulation by another hepatic lectin[9]. In animal cells transfected with an expression vector containing a cDNA copy, the polypeptide produced will subsequently be modified by glycosyltransferases present in the host cells. It is critical, therefore, that the host cells synthesize carbohydrate moieties similar to those naturally produced. The therapeutic use of recombinant glycoprotein hormones will depend on their mimicking the actions of naturally produced hormones, including the rate of clearance from the circulatory system.

Molecular Biology of Erythropoiesis
Edited by J. L. Ascensao *et al.*
Plenum Press, New York

53

Recently, we have analyzed the carbohydrate structure of recombinant erythropoietin produced in Chinese hamster ovary cells that were transfected with a human erythropoietin cDNA[1,10]. In addition, we have studied the clearance of this recombinant erythropoietin in relation to its carbohydrate structure[11]. In this article, we summarize these studies, focusing on the role of carbohydrates with respect to the clearance of the recombinant erythropoietin from the circulation.

Isolation of recombinant and urinary erythropoietin

Chinese hamster ovary cells were transfected with an expression vector that contains human erythropoietin cDNA. Erythropoietin was purified from the spent medium of those cells as described previously[7]. The purification procedure was modified from that of Miyake et al.[12], and fractionation by a reverse-phase high-performance liquid chromatography was included[7]. Erythropoietin was also purified from the urine of aplastic anemia patients according to Miyake et al.[12], with a similar modification. These erythropoietin samples were prepared by Chugai Pharmaceuticals, Ltd. (Tokyo) (see Fig. 1).

Structure of erythropoietin polypeptide

The peptide moiety of erythropoietin consists of 165 amino acid residues, including four cysteine residues[13,14]. These four cysteine residues are connected to form disulfide loops as shown in Fig. 2. The amino acid sequence and the analysis of glycopeptides indicate that asparagine-- linked saccharides are attached to the residues 24, 38, and 83, while 0-glycan is attached to residue 126[10,13]. By comparison of erythropoietin from different species, the secondary structure of human erythropoietin can be depicted as shown in Fig. 2[15]. This structure clearly shows that

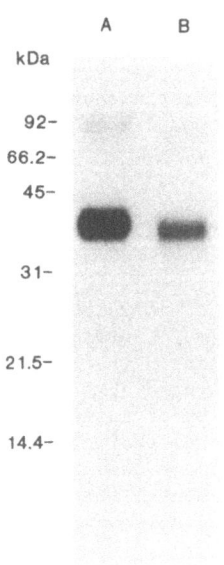

Fig. 1. Autoradiogram of SDS-gel electrophoresis of recombinant erythropoietin (A) and urinary erythropoietin (B). Purified erythropoietins were iodinated with ^{125}I according to Greenwood et al[27]. The radioactively labeled proteins were then applied to SDS-polyacrylamide gel electrophoresis (gel concentration 15%) according to Laemmli[28], and the gel was directly autoradiographed with Kodak X-ray AR-5 film (from ref. 1).

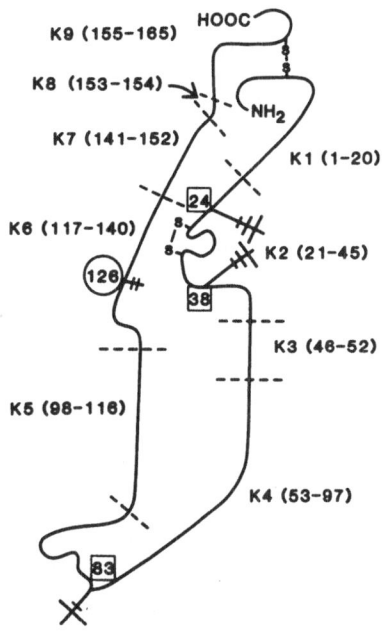

Fig. 2. Predicted structure of human
erythropoietin. (Adopted
from references 14 and 15). K1
to K9 are peptides generated
by Endo Lys-C and the amino
acid residue numbers are indicated
in the parentheses. The amino
acid residue 24, 38 and 83 attached N-linked saccharides,
whereas the residue 126 attaches
0-linked saccharides

two glycosylation sites at residue 24, 38 and the third site at residue
83 are in a quite different environment.

The most recent results indicate that the poplypeptide of erythropoietin
is processed to lose the last amino acid, argiane at residue 166. This
processing is naturally taking place in both recombinant erythropoietin
and erythropoietin obtained from urine. The removal of this amino acid
does not change erythropoietin activity[14].

Structural analysis of saccharides attached to recombinant as well as
urinary erythropoietin

Carbohydrate chains attached to recombinant and urinary erythropoietin
were released by N-glycanase and fractionated by high-performance liquid
chromatography. The structures of fractionated saccharides were determined
by methylation analysis before and after treatment with specific glycosidases
and fast atom bombardment-mass spectrometry. Both erythropoietins were
found to contain one 0-linked oligosaccharide/mol of the proteins, and
its major component was elucidated to be NeuNAc $\alpha 2 \rightarrow 3$ and $\beta 1 \rightarrow 3$ (NeuNAc
$\alpha 2 \rightarrow 6$) GalNAc in both proteins. Both proteins contain three N-linked
saccharides, and they were found to mainly consist of sialylated tetra--
antennary saccharides of which a portion contains polylactosaminyl repeats
(Table I, Fig. 3). All of these saccharides are sialylated by $\alpha 2 \rightarrow 3$

Table 1. Structures of asialo N-linked saccharides obtained from human recombinant erythropoietin. The structures are based on the previous work (Sasaki et al., 1987).

Name	Structure[a]	N-24, N-38, N-83

Bi-antennary

```
                                                      Fuc
                                                      α1
                                                      ↓
    Galβ1→4GlcNAcβ1→2Manα1                            6
                          6
                            Manβ1→4GlcNAcβ1→4GlcNAcOH        6%   0   0
                          3
    Galβ1→4GlcNAcβ1→2Manα1
```

Tri-antennary

```
    Galβ1→4GlcNAcβ1→2Manα1                      Fuc
                                                α1
                                                ↓
    Galβ1→4GlcNacβ1          6                  6
                    4          Manβ1→4GlcNAcβ1→4GlcNAcOH
                     Manα1   3
                    2
    Galβ1→4GlcNAcβ1


    Galβ1→4GlcNAcβ1                             Fuc
                  6                             α1
                   Manα1                        ↓
                  2      6                      6
    Galβ1→4GlcNAcβ1       Manβ1→4GlcNAcβ1→4GlcNAcOH
                        3
    Galβ1→4GlcNAcβ1→2Manα1
```

(bracketed group) 15%   5%   6%

Tetra-antennary[b]

```
    Galβ1→4GlcNAcβ1
                  6
                   Manα1                        Fuc
                  2                             α1
    Galβ1→4GlcNAcβ1      6                      ↓
                          Manβ1→4GlcNAcβ1→4GlcNAcOH    45%  49%  85%
    Galβ1→4GlcNAcβ1      3
                  4    /
                   Manα1
                  2
    Galβ1→4GlcNAcβ1
```

Tetra-antennary Lac$_1$

```
    Galβ1→4GlcNAcβ1→3Galβ1→4GlcNAcβ1
                                    6
                                     Manα1                Fuc
                                    2                     α1
    Galβ1→4GlcNAcβ1                      6                ↓
                                          Manβ1→4GlcNAcβ1→4GlcNAcOH
    Galβ1→4GlcNAcβ1                      3
                  4    /
                   Manα1
                  2
    Galβ1→4GlcNAcβ1


    Galβ1→4GlcNAcβ1
                  6
                   Manα1                                  Fuc
                  2                                       α1
    Galβ1→4GlcNAcβ1→Galβ1→4GlcNAcβ1      6                ↓
                                          Manβ1→4GlcNAcβ1→4GlcNAcOH
    Galβ1→4GlcNAcβ1                      3
                  4    /
                   Manα1
                  2
    Galβ1→4GlcNAcβ1


    Galβ1→4GlcNAcβ1
                  6
                   Manα1                                  Fuc
                  2                                       α1
    Galβ1→4GlcNAcβ1      6                                ↓
                          Manβ1→4GlcNAcβ1→4GlcNAcOH
    Galβ1→4GlcNAcβ1→3Galβ1→4GlcNAcβ1     3
                                    4  /
                                     Manα1
                                    2
                    Galβ1→4GlcNAcβ1
```

(bracketed group) 25% 32% 9%

Table 1. Continued

Name	Structure[a]	N-24, N-38, N-83

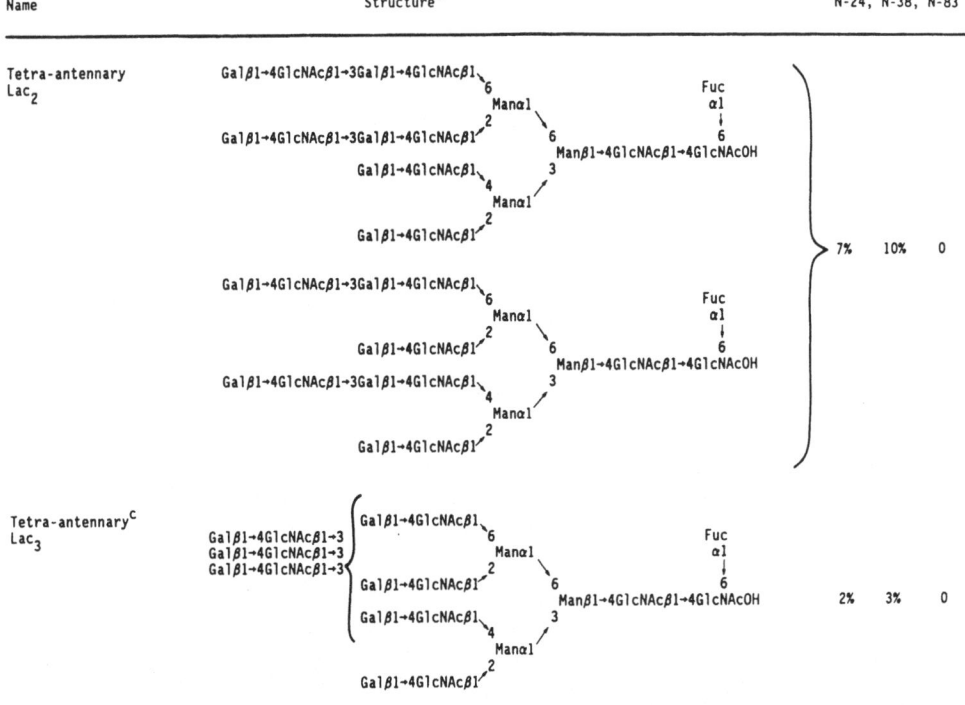

a   15% of the saccharides lack fucose attached to reducing terminal N-acetylglucosamine
b   Very small amount of tri-antennary, Lac₁ is also present in this fraction.
c   A small proportion of this saccharide contains (Galβ1→4GlcNAcβ1→3)₃ as a side chain.

linkages. The results also showed that the carbohydrate moiety of urinary erythropoietin is indistinguishable from recombinant erythropoietin, except for a slight difference in sialylation (see ref. 1,16).

Since erythropoietin contains only three N-linked saccharides, it is a good model for analyzing saccharide structures that are attached at different sites. Glycopeptides were separated by reverse-phase HPLC and those containing only one carbohydrate chain were isolated. The carbohydrate moiety was released from each glycopeptide fraction and analyzed by HPLC. The results are summarized as follows (Table I, ref. 10). Saccharides at Asn^{24} are heterogenous and consist of bi-antennary, tri-antennary, and tetra-antennary saccharides with or without N-acetyllacto-saminyl repeats; (2) saccharides at Asn^{38} mainly consist of well-processed saccharides, such as tetra-antennary saccharides with or without N-acetyllac-tosaminyl repeats; (3) saccharides at Asn^{83} are homogenous in the backbone structure and are composed mainly of tetra-antennary without N-acetyllacto-saminyl repeats. These results clearly indicate that the protein structures and, possibly, the carbohydrate chain at the neighboring site greatly influences glycosylation of a given glycosylation site (ref. 10).

Clearance of recombinant erythropoietin from the plasma circulation of rats

In order to evaluate the role of carbohydrates in the survival rate of erythropoietin in the plasma, recombinant erythropoietin was labeled with [^{125}I] and injected into rats intravenously, i.v. A portion of

A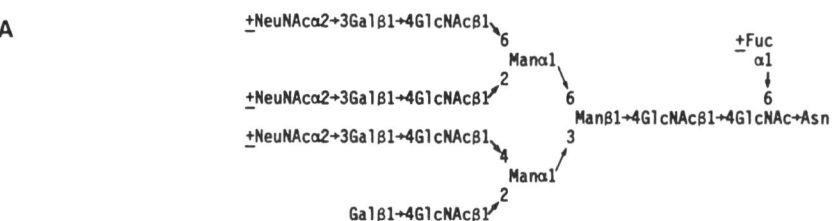

B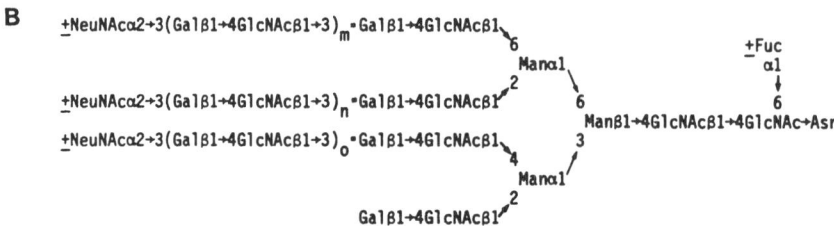

Fig. 3. Proposed structures of tetra-antennary saccharides (A) and
tetra-antennary saccharides with N-acetyllactosaminyl repeats (B) obtained from recombinant erythropoietin.
A. Sialylation takes place preferentially at the side chain
arising from C-6, then from C-2 of 2,6-substituted mannose
and c-4 of 2,4-substituted mannose.
B. The saccharides with one N-acetyllactosaminyl repeat;
m=1, n=0, o=0, in 90% of the molecules and o=1, m=n=0 in
10% of the molecules. The saccharides with two N-acetyllactosaminyl repeats: m=n=1, o=0 in 60% of the molecules,
m=o=1 and n=0 in 40% of the molecules. The saccharides
with three N-acetyllactosaminyl repeats are m+n+o=3.
(from ref. 1).

the intact erythropoietin was cleared rapidly, but this apparent clearance
is most likely due to the specific binding of erythropoietin to bone
marrow cells, since this apparent early-phase clearance disappears when
a large quantity (more than 4 pmole) of erythropoietin is injected. The
data suggest that the bone marrow cells are under-saturated with erythropoietin, and a part of the exogenous erythropoietin is quickly taken
up by bone marrow cells. The majority of the recombinant erythropoietin
remained in circulation relatively stable, up to 30 minutes. This longer
β-phase has a half-life of 108 minutes (Fig. 5).

To determine the role of terminal sialic acid in the clearance of
recombinant erythropoietin, similar experiments were performed after
sialic acid residues were enzymatically removed from the recombinant
erythropoietin. In contrast to the intact erythropoietin, almost all
recombinant asialo-erythropoietin rapidly disappeared from plasma within
6 minutes. It was noticed that the radioactivity in plasma increased
gradually between 10 and 30 minutes. In order to elucidate the nature
of this radioactive component, plasma components collected during clearance
experiments were analyzed by sodium dodecyl sulfate-polyacrylamide gel
electrophasis. While the recombinant erythropoietin remained intact
up to 30 minutes (Fig. 6A), the asialo-erythropoietin disappeared quickly
within 6 minutes of injection. However, we did not detect any specific
polypeptide band that corresponded to the increase in total serum activity
found between the 10- and 30-minutes time (Fig. 6B). In order to test
whether the increase in radioactivity is due to degration of erythropoietin,
serum samples of labeled asialo-erythropoietin collected one minute and
30 minutes after injection were analyzed by Sephadex G-50 gel filtration.

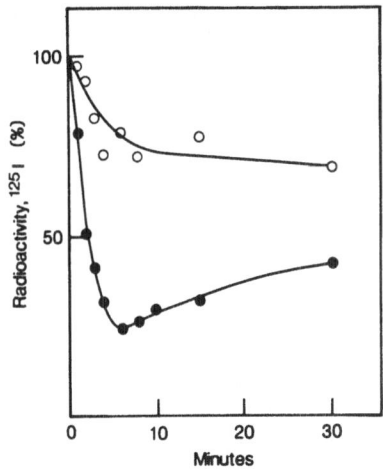

Fig. 4. Clearance of intact and
desialyzed erythropoietins
from circulation of rats.
]125I]-labeled recombinant
erythropoietin was injected
intravenously into rat before
(-O-) and after (-●-) siali-
dase treatment and radio-
activity in blood was measured
at the time intervals shown.
(from ref. 11).

The majority of radioactivity from the sample collected one minute after
injection was eluted at the void volume, as should be expected from an
undegraded glycoprotein. The sample collected 30 minutes after injection
was eluted at the column volume from the same column, thus indicating
that the radioactivity increase between 10 and 30 minutes is due to the
release of degraded erythropoietin, which are too small to be detected
by SDS-polyacrylamide gel electrophoresis. Similar results were reported
by Spink and Hogans[17].

Uptake of asialo-erythropoietin to liver

In order to determine where asialo-erythropoietin was taken up,
the radioactivity present with the four major organs of the rat were
analyzed 30 minutes after injection. As shown in Fig. 7A, the predominance
(85%) of asialo-erythropoietin was taken up by the liver, probably by
receptor-mediated endocytosis specific to galactose in clearance (see
below). The distribution of intact erythropoietin in organs was also
determined 30 minutes or 3 hours after injection. At 30 minutes after
injection, the relative amount accumulated in the liver is less than
that for the asialo form. Three hours after injection, the radioactivity
taken up in the kidney increased significantly (Fig. 7C), and this label
may well be excreted (urinary) material.

The effect of N-acetyllactosamine repeats in the clearance of erythro-
poietin

The major carbohydrates of recombinant erythropoietin are tetra-antennary saccharides[1,10,16]. Among these saccharides, some contain N-acetyllactosamine repeats which are attached to the tetraantennay core saccharides.

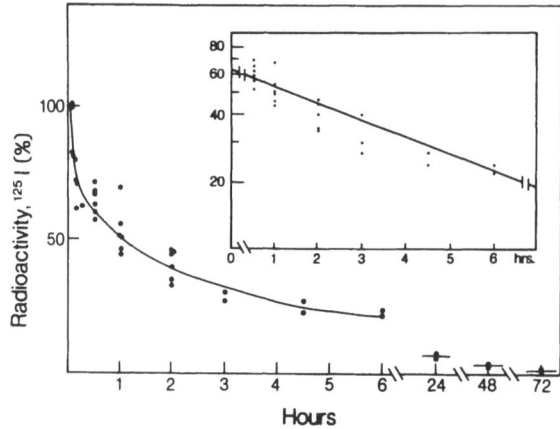

Fig. 5. Clearance of intact recombinanty erythropoietin from plasma circulation over a
period of 72 hours. Results obtained
for six rats are presented. (from ref. 11(.

Oligosaccharides with one N-acetyllactosamine repeat represent 32.1%
of the total saccharides; those with two N-acetyllactosamine repeats
represent 16.5%; and those with three N-acetyllactosamine repeats, 4.7%[1,10].
In order to evaluate the role of N-acetyllactosamine repeats on clearance
of glycoproteins, erythropoietin was fractionated by tomato-lectin-Sepharose,
which binds carbohydrates with more than three N-acetyllactosamine repeats.
The clearance rate of the tomato lectin-bound (about 20% of total erythropoietin) and unbound factions was tested by similar experiments described
above. The results clearly indicate that the erythropoietin bound to
tomato lectin (containing N-acetyllactosamine repeats) was cleared more
rapidly than the unbound fraction (Fig. 8A). As shown in Fig. 7D, the
erythropoietin bound to tomato lectin was taken up by organs in a similar
way to the uptake for total intact erythropoietin, except that the uptake
by the liver is slightly higher in the tomato lectin-bound fraction.

Erythropoietin was also fractionated by Datura stramonium lectin,
which binds tetra-antennary saccharides. However, there was no difference
in the clearance rate between Datura-bound and unbound components of
recombinant erythropoietin (Fig. 8B).

These results suggest that saccharides containing N-acetyllactosamine
repeats can be cleared more efficiently. In order to determine whether
this applies to polylactosaminoglycans in general, polylactosaminoglycans
isolated from erythrocytes, which have bi-antennary structure with polylactosaminyl side chains, were tested[18,19]. These glycopeptides were rapidly
cleared from blood circulation, and more than 90% of the radioactivity
was taken up in the liver. This clearance pattern was identical regardless
of whether the erythrocytes polylactosamines were intact or disialylated
or whether they were linear or branched (Fig. 9).

In order to know whether the uptake of polylactosamine is through
a receptor other than galactose-binding proteins, competitors were added
to the assay. Asialo-α_1-acid glycoprotein released polylactosamines
in the early stage of uptake, whereas intact α_1-acid glycoprotein[20] or
lactose had no effect. The results suggest that the N-acetyllactosamine
repeats in intact α_1-acid glycoprotein is not enough to release erythrocyte
polylactosaminoglycan, but that the galactose terminal residues exposed
by desialylation of α_1-acid glycoprotien can compete with the polylacto-

A

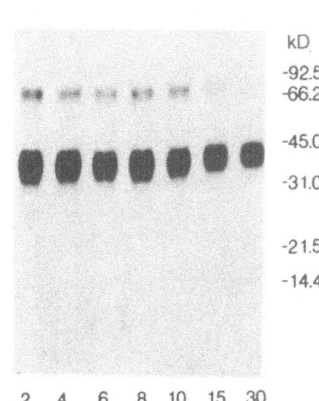

kD
-92.5
-66.2

-45.0

-31.0

-21.5

-14.4

2 4 6 8 10 15 30

B

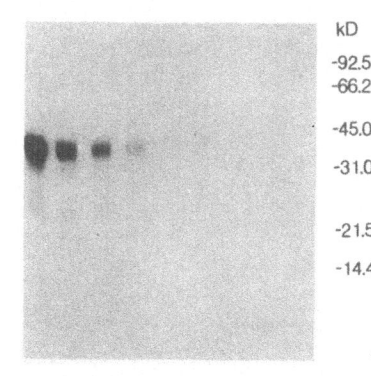

kD
-92.5
-66.2

-45.0

-31.0

-21.5

-14.4

1 2 3 4 6 8 10 15 30

Fig. 6.   SDS-polyacrylamide gel
electrophoresis of intact
and desialyzed recombinant
erythropoietin recovered
from rat plasma. Plasma
was applied to SDS-poly-
acrylamide gel electro-
phoresis and [^{125}I] was
detected by autoradio-
graphy. 14% polyacrylamide
was used.

A.   Intact erythropoietin
B.   Desialyzed erythropoietin.

Numbers at the bottom of each
lane show the time (min.) when
plasma was taken after the injection of the protein. (from
ref. 11).

samines. These observations suggest that a glycoprotein with a number
of N-actyllactosaminyl repeats are probably recognized by the same galactose
binding protein on hepatocytes, even when the glycoprotein is sialylated.

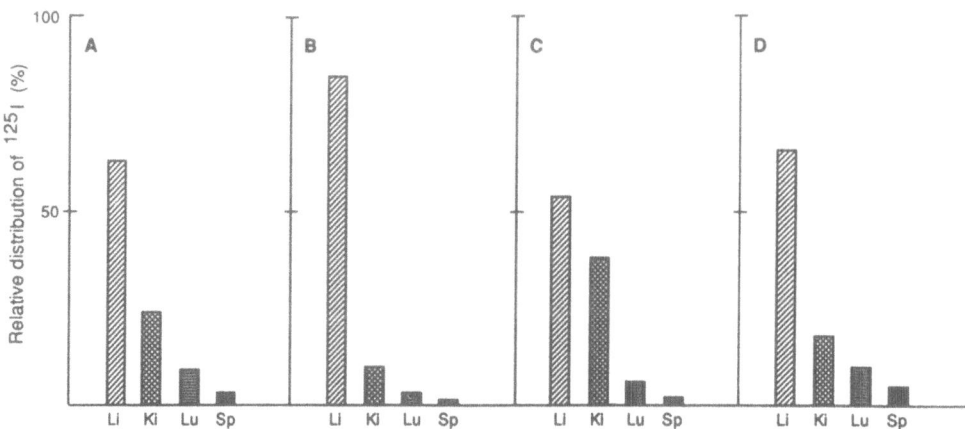

Fig. 7. Histograms showing the distribution of $[^{125}I]$-erythropoietin
among organs. The relative radioactivity incorporated into
the liver (Li), kidney (Ki), lung (Lu) and spleen (Sp) is
shown.

A. Intact erythropoietin at 30 minutes.
B. Asialo erythropoietin at 30 minutes.
C. Intact erythropoietin at 3 hours.
D. Tomato lectin-bound erythropoietin at 30 minutes.

DISCUSSION

The present results demonstrate that the carbohydrate structure
present in recombinant erythropoietin is indistinguishable from that
of erythropoietin of natural origin. Since the recombinant erythropoietin
was made in Chinese hamster ovary cells, it is apparent that the Chinese
hamster ovary cells and human kidney cells which produce natural erythropoietin have a similar set of glycosyltransferases. This similarity
is extremely critical since different carbohydrates could provoke immune
reaction toward recombinant erythropoietin in blood circulation.

Genetic engineering technology has enabled the production of a large
quantity of cytokines such as interferons, interleukin-2, and tumor necrosis
factor[23]. Many of these recombinant cytokines, however, have been found
to be unstable in vivo or cause side effects upon administration to animals
and humans[23]. Fortunately, erythropoietin produced by transfection of
animal cells has been greatly successful for the treatment of patients[24,25].
The success of clinical trials of erythropoietin is most likely due to
the fact that the erythropoietin targets narrowly specific cells such
as erythroid progenitor cells. However, another factor contributing
to its biologic activity in vivo is the stability of the erythropoietin
molecule. We have shown that stability is largely dependent on a high
proportion of carbohydrates of the recombinant erythropoietin being in
the correct form to escape detection and removal by galactose binding
receptors of hepatocytes. The determination of the correctly glycosylated
forms of potential recombinant therapeutic agents and selection of proper
expression vectors and hosts to elicit those structures may be very important
in future studies.

The present results showed that the sialylated N-glycan modifications
are required for the survival of recombinant erythropoietin. Once sialic
acid is removed, tri-antennary and tetra-antennary N-linked saccharides
are taken up by hepatic lectin[21] so that glycoprotein such as erythropoietin,

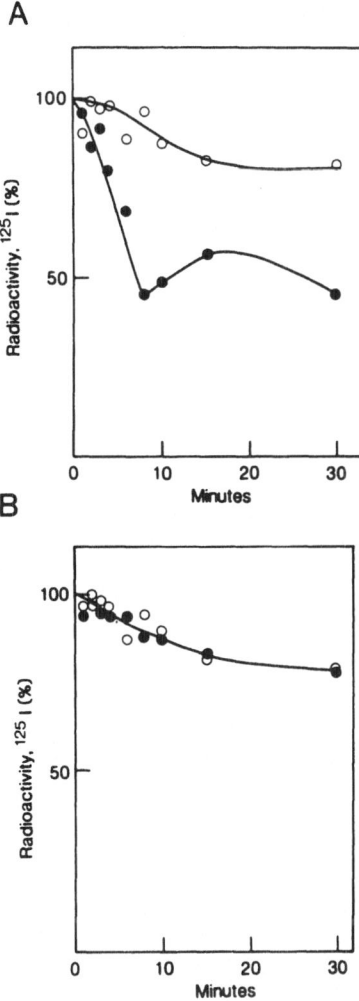

Fig. 8. A.   Clearance of erythropoietin
             separated by tomato lectin.

        -O- Tomato unbound fraction
             (erythropoietin without
             lactosamine repeats or
             with less than three
             lactosamine repeats).

        -●- Tomato bound fraction
             (erythropoietin with three
             or more lactosamine
             repeats).

     B.   Clearance of erythropoietin
          separated by Datura agglutin.

        -O- Datura lectin-unbound
             fraction.

        -●- Datura lectin-bound faction.

     (from ref. 11).

63

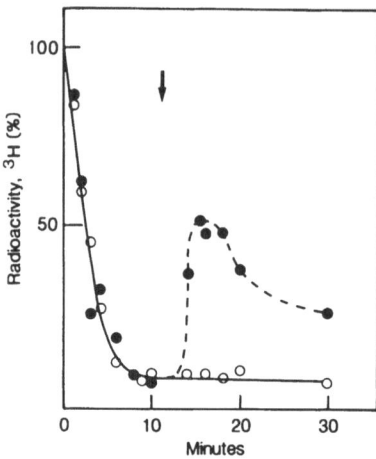

Fig. 9. Clearance of polylactosaminoglycans.
[³H]-labeled branched polylactosaminoglycans isolated from erythrocytes (see the text) were injected
into the rat. At the time shown by the
arrow, asialo-α_1-acid glycoprotein (●)
or lactose (5 mg in 0.2 ml, O) was injected IV. Linear polylactosaminoglycans gave similar results (not
shown). (from ref. 11).

which has those structures, will be taken up quickly by the liver after
desialylation. Since erythropoietin produced by Chinese hamster ovary
cells contains only $\alpha2\rightarrow3$-linked sialic acid, some side chains of carbohydrates lack sialic acid. It will be interesting to know if further sialylation by $\alpha2\rightarrow6$ sialyltransferase will lengthen the survival rate of erythropoietin.

The behavior of polylactosamines in plasma circulation is consistent
with the fact that polylactosamines are abundant in plasma membranes
but rare in serum. As exemplified by the band 3 glycoprotein in erythrocyte
membranes[18,19], the polylactosaminoglycans are major glycoconjugates
in some cells in the human body. Little is known, however, about the
catabolism of polylactosamines. The present results suggest that polylactosamines could be rapidly internalized by hepatocytes if they were released
to plasma. The presence of keratan sulfate, a glycosaminoglycan consisting
of sulfated polylactosamines, in human serum[22] suggests that the sulfation
at C-6 on galactose residues protects the polylactosamines from recognition
by the galactose-binding lectin and subsequent clearance by hepatocytes.

In contrast to sialylation, which prevents rapid clearance from
blood circulation, the addition of N-acetyllactosamine repeats facilitates
the uptake of glycoproteins by the liver through hepatic galactose-binding
protein (Fig. 10). It is important to see if erythropoietin, which does
not contain N-acetyllactosamine repeats, can be obtained by choosing cells
which lack glycosyltransferases[26] which are responsible for polylactosamine
synthesis. It will then be interesting to determine if erythropoietin
without N-acetyllactosamine repeats survive longer in the blood circulation.

Along the same line, it would be interesting to choose cell lines
for transfection, which lack in N-acetylglucosamine transferases responsible
for tri-or tetra-antennary saccharides. When sialic acid residues are

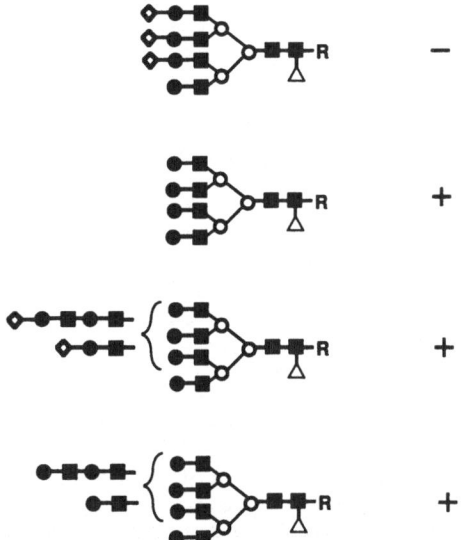

Fig. 10.   The structure of the carbohydrate moiety
of recombinant erythropoietin and its
binding to hepatic galactose-binding
receptor.   Structures of carbohydrates are
adopted from the previous studies[1,10].
From the top, sialylated tetra-antennary,
asialo tetra-antennary, sialylated tetra-
antennary with three lactosamine units,
and asialo-tetra-antennary with three
lactosamine units.   The carbohydrate that
binds to the hepatocytes (+ in this
figure) is cleared quickly from the
circulation. ◺ = sialic acid;   ○ =
galactose; ■ = N-acetylglucosamine;
○ = mannose;   Δ = fucose.   (from ref. 11).

removed from blood circulation, glycoproteins containing tri- or tetra--
antennary saccharides are expected to be taken up by the liver[21].   Then
it would be possible that glycoproteins, containing only bi-antennary
saccharides, have very sustained life-time in the blood circulation because
asialo bi-antennary saccharides are not recognized by hepatic lectin.
If the saccharides are fully processed and no high-mannose forms remained,
those recombinant glycoproteins should have a much longer effective concentration in blood circulation. We believe that the attempts to obtain such
glycoproteins are worthwhile.

ACKNOWLEDGMENTS

    The authors thank Dr. Tsutomu Kawaguchi of Chugai Pharmaceuticals
for his initiation of this joint project, and Ms. Henny Bierhuizen and
Kaarin Soma for secretarial assistance.   The work carried out in our
laboratories was supported by grants CA 33000 (to M.F.) and DK 37016
(to M.N.F.).

REFERENCES

1. Sasaki, H., B. Bothner, A. Dell, and M. Fukuda. 1987. Carbohydrate structure of erythropoietin expressed in Chinese hamster ovary cells by a human erythropoietin cDNA. J. Biol. Chem. 262: 12059-12076.

2. Lai, P.H., R. Everett, F.F. Wang, T. Arakawa, and E. Goldwasser. Structural characterization of human erythropoietin. J. Biol. Chem. 261: 3116-3121.

3. Lowry, P.H., G. Keighley, and H. Borsook. 1960. Inactivation of erythropoietin by neuraminidase and mild substitution reactions. Nature 185: 102-103.

4. Goldwasser, E., and C.K-H. Kung. 1968. Part II. Chemistry and purification of erythropoietin: Progress in the purification of erythropoietin. Ann. NY Accad. Sci. 149: 49-53.

5. Goldwasser, E., C.K-H. Kung, and J. Eliason. 1974. On the mechanism of erythropoietin-induced differentiation XIII. The role of sialic acid in erythropoietin action. J. Biol. Chem. 249: 4202-4206.

6. Morell, A.G., K.A. Irvine, I. Sternlieb, I.H. Scheinberg, and G. Ashwell. 1968. Physical and chemical studies on ceruloplasmin: V. Metabolic studies on sialic acid-free ceruloplasmin in vivo. J. Biol. Chem. 243: 155-159.

7. Jacobs, K., C. Shoemaker, R. Rudersdorf, S.D. Neill, R.M. Kaufman, A. Mufson, J. Seehra, S.S. Jones, R. Hewick, E.F. Ritch, M. Kawakita, T. Shimizu, and T. Miyake. 1985. Isolation and characterization of genomic and cDNA clones of human erythropoietin. Nature 313: 806-810.

8. Lin, F-K., S. Suggs, C-H. Lin, J.K. Browne, R. Smalling, J.C. Egrie, K.K. Chen, G.M. Fox, F. Martin, Z. Stabinsky, S.M. Badrawi, P-H Lai, and E. Goldwasser. 1985. Cloning and expression of the human erythropoietin gene. Proc. Natl. Acad. Sci. USA 82: 7580-7584.

9. Mizuno, Y., Y. Kozutsumi, T. Kawasaki, and I. Yamashina. 1981. Isolation and characterization of a mannan-binding protein from rat liver. J. Biol. Chem. 256: 4247-4252.

10. Sasaki, H., N. Ochi, A. Dell, and M. Fukuda. 1988. Site-specific glycosylation of human recombinant erythropoietin: Analysis of glycopeptides or peptides at each glycosylation site by fast atom barbardment-mass spectrometry. Biochemistry 27: 8618-8626.

11. Fukuda, M.N., H. Sasaki, L. Lopez, and M. Fukuda. 1989. Survival of recombinant erythropoietin in the circulation: The role of carbohydrates. Blood 73: 84-89.

12. Miyake, T., C. K-H. Kung, and E. Goldwasser. 1977. Purification of human erythropoietin. J. Biol. Chem. 252: 5558-5564.

13. Lai, P-H., R. Everett, F-F. Wang, T. Arakawa, and E. Goldwasser. 1986. Structural characterization of human erythropoietin. J. Biol. Chem. 261: 3116-3121.

14. Recny, M.A., H.A. Scoble, and Y. Kim. 1987. Structural characterization of natural human urinary and recombinant DNA-derived erythropoietin: Identification of des-arginine 166 erythropoietin. J. Biol. Chem. 262: 17156-17163.

15. McDonald, J.D., F-K. Lin, and G. Goldwasser. 1986. Cloning, sequencing, and evolutionary analysis of the mouse erythropoietin gene. Mol. Cell. Biol. 6: 842-848.

16. Takeuchi, M., S. Takasaki, H. Miyazaki, T. Kato, S. Hoshi, N. Kochibe, and A. Kobata. 1988. Comparative study of the asparagine-linked sugar chains of human erythropoietins purified from urine and the culture medium of recombinant Chinese hamster ovary cells. J. Biol. Chem. 263: 3657-3663.

17. Spivak, J.L., and B.B. Hogans. 1989. The in vivo metabolism of recombinant human erythropoietin in the rat. Blood 73: 90-99.

18. Fukuda, M., A. Dell, and M.N. Fukuda. 1984. Structure of fetal lactosaminoglycan: The carbohydrate moiety of band 3 isolated from human umbilical cord erythrocytes. J. Biol. Chem. 259: 4782-4791.

19. Fukuda, M., A. Dell, J.E. Oates, and M.N. Fukuda. 1984. Structure of branched lactosaminoglycan, the carbohydrate moiety of band 3 isolated from adult human erythrocytes. J. Biol. Chem. 259: 8260-8273.

20. Yoshima, H., A. Matsumoto, T. Mizuochi, T. Kawasaki, and A. Kobata. 1981. Comparative study of the carbohydrate moiety of rat and human plasma α_1-acid glycoprotein. J. Biol. Chem. 256: 8476-8484.

21. Baenziger, J., and D. Fiete. 1980. Galactose and N-acetylgalacto-samine-specific endocytosis of glycopeptides by isolated rat hepatocytes. Cell 22: 611-620.

22. Thonar, E., R.F. Meyer, R.F. Dennis, M.E. Lenz, B. Maldonadao, J.R. Hassell, A.T. Hewitt, W.J. Stark, E.L. Stock, K.E. Kuettner, and G.K. Klintworth. 1986. Absence of normal keratansulfate in the blood of patients with macular corneal dystrophy. Am. J. Opthalmol. 102: 561.

23. Moertel, C.G. 1986. On lymphokines, cytokines and breakthroughs. JAMA 256: 3141, (editorial).

24. Eschbach, J.W., J.C. Eggrie, M.R. Downing, J.K. Browne, and J.W. Adamson. 1987. Correction of the anemia of end-stage renal disease with recombinant human erythropoietin: Results of a phase I and II clinical trial. N. Engl. J. Med. 316: 73.

25. Winearls, C.G., D.O. Oliver, M.J. Pippard, C. Reid, M.R. Downing, and P.M. Cotes. 1986. Effect of human erythropoietin derived from recombinant DNA on the anemia of patients maintained by chronic haemodialysis. Lancet 2: 1175.

26. Piller, F., and J-P. Cartron. 1983. FUDP-GlcNAc:Galβ1-4Glc(NAc)β1-3N-acetylglucosaminyltransferase indentification and characteri-zation in human serum. J. Biol. Chem. 258: 12293-12299.

27. Greenwood, F.C., W.M. Hunter, and J.S. Glove. 1963. The preparation of ^{123}I-labelled human growth hormone of high specific radio-activity. Biochem. J. 89: 114-123.

28. Laemmli, U.K. 1970. Cleavage ot structural proteins during assembly of the head of bacteriophage T4. Nature 227: 680-685.

CELLULAR LOCALIZATION OF ERYTHROPOIETIN GENE TRANSCRIPTION

Catherine Lacombe, Jean-Louis Da Silva*, Patrick Bruneval*,
Jean-Loup Salzmann*, Nicole Casadevall, Jean-Pierre Camilleri*,
Jean Bariety*, Bruno Varet, Pierre Tambourin

INSERM U152, Hopital Cochin
*INSERM U28, Hopital Broussais
Paris, France

ABSTRACT

Erythropoietin producing cells were identified in the murine hypoxic
kidney by in situ hybridization. The positive cells were peritubular
cells, most likely endothelial cells of the cortex and outer medulla.
Glomerular and tubular cells were not labelled.

In three patients with renal adenocarcinomas associated with polycythemia, a strong Epo message was observed on Northern blot analysis.
Using in situ hybridization, a strong labelling was observed in all cases
on the tumor cells which are of tubular origin.

Cellular localization of erythropoietin gene transcription

Erythropoietin (Epo) is the hormone that controls the red blood
cell production in mammals. Epo synthesis is regulated via feedback
mechanisms involving tissue oxygen tension. Since the work of Jacobson
et al.[1], the kidney is known to be the major site of Epo production (for
review see[2]). Within the kidney, the identity of the cells synthesizing
Epo was controversial. Immunofluorescence data[3], and glomerular[4] and
mesangial[5] culture studies supported a glomerular nature of renal Epo-producing cells. On the contrary, an extraglomerular nature of Epo-producing
cells, i.e. tubular or interstitial, was suggested by studies based on
renal tissue fractions[6,7]. Using in situ hybridization technique, we
recently demonstrated that the renal Epo-producing cells were peritubular
cells in the murine hypoxic kidney cortex[8]. We also studied, 3 human
renal adenocarcinomas associated with polycythemia, and reported that,
in this malignant situation, tumor cells of epithelial origin were the
site of Epo production[9].

RESULTS AND DISCUSSION

Erythropoietin mRNA expression in organs of hypoxic mice

On a Northern blot sequentially hybridized with 1) a mouse Epo probe,
2) and 3) two housekeeping gene probes to assess that similar amounts

of poly(A)+ RNAs were loaded, a strong 1.8 kb Epo signal was detected
in the kidneys of anemic mice, and to a lesser extent in the anemic liver
(Fig. 1.). This Epo mRNA size was identical to the murine Epo mRNA previously described[10,11]. Other organs (brain, testis, salivary glands) were
all negative even after a longer exposure of the film (10 days). In
normal mouse kidneys, there was a very faint Epo signal showing that
there was a basal mRNA expression in this organ in non anemic mice.

Identification of Epo-producing cells in mouse hypoxic kidney

In situ hybridization of anemic mouse kidney sections, using a ^{35}S-
labelled 243 bp Epo probe[12], detected an intense signal in many cells
of the renal cortex and the outer medulla. The positive cells were clearly
in a peritubular location, out of the glomeruli and out of the tubules
(Fig. 2.). In the inner medulla, no signal was detected. Control experiments, including ribonuclease treated sections, and sections hybridized
with pUC 18 plasmid were negative. In kidney sections of non anemic
mice, in situ hybridization with the Epo probe did not detect any signal.
To confirm these results, we submitted hybridized sections to image analysis.
We used a computer vision image processor (NS 1500, Nachet-Vision, France).
Silver grains were isolated by "Top hat" transformation, to select them
according to their size and contrast[13]. Because of the silver grain
image overlapping, their surface rather than their absolute number was
calculated. For silver grains localization, the operator studied successive-
ly the cells of the glomeruli, the tubules, and the peritubular areas.
The results were referred to a unit area of 100 μm^2. As shown on Fig.
3, the silver grains counted on the glomeruli and the tubules were at
the background level. In contrast, cortical and outer medullary peritubular
cells of the anemic mouse kidney were strongly labelled.

To further characterize these peritubular cells we used antibodies
directed against von Willebrand factor and against the murine monocyte
macrophage specific antigen F4/80[14]. Kidney sections of hypoxic and
control mice had similar immunolabelling pattern. Anti F4/80 antibody,
which strongly reacted with liver Kupffer cells did not react with cortical
peritubular cells, indicating that no macrophages reacting with the antibody

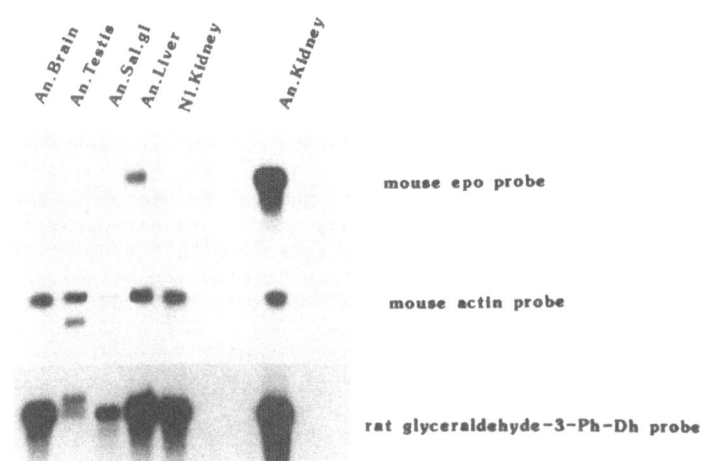

Fig. 1. Northern blot of poly (A)$^+$ RNAs extracted from different organs of anemic mice and from normal mouse
kidney. 5 µg of poly (A)$^+$ RNAs were electrophoresed.
Hybridizations were sequentially performed with
1) mouse Epo probe; 2) mouse actin probe; 3) rat
glyceraldehyde-3-Phosphate-Deshydrogenase probe.

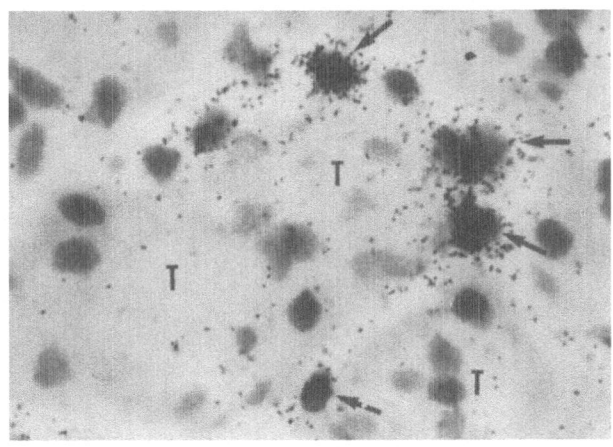

Fig. 2.   In situ hybridization of anemic mouse kidney
          section using a ^{35}S labelled mouse Epo probe;
          Positive cells are in a peritubular location
          (arrows), the tubular cells are negative (T)
          (X800).

were present within the hybridization areas. These data differed from
previous report where bone marrow macrophages have been described to
produce Epo in culture[15].

     In contrast, using an anti von Willebrand factor antibody, a strong
positivity was observed in endothelial cells, lining the peritubular
capillaries where most silver grains were seen. Furthermore, an ultra-
structural study showed that most of the peritubular cells in the renal
cortex and the outer medulla were capillary endothelial cells.

     To further demonstrate that peritubular cells which produce Epo
are endothelial cells, double labelling experiments with, on the one
hand an anti von Willebrand factor antibody and on the other hand the
[35]S labelled murine Epo probe are in progress.

Identification of Epo-producing cells in human renal adenocarcinoma associated with polycythemia

     In three patients, a strong Epo expression was observed in the tumor
RNA extracts whereas no Epo message could be detected in the non-tumoral
adjacent renal tissues. Using in situ hybridization, and a ^{35}S labelled
monkey Epo probe[16], numerous silver grains were found on the tumor cells.
The stromal and endothelial cells were at the background level (Fig. 4).
These tumor cells from renal adenocarcinomas were of epithelial nature
as demonstrated by a strong cytoplasmic labelling with an anti cytokeratin
antibody. They are currently supposed to derive from proximal tubular
cells[17].

CONCLUSION

     These latter results were in apparent contradiction with those describ-
ing the site of Epo production in murine hypoxic kidney[8,18]. The following
hypotheses can be discussed to explain these discrepancies:

     The first hypothesis would be that the malignant transformation

Silver grains area / 100 μm^2

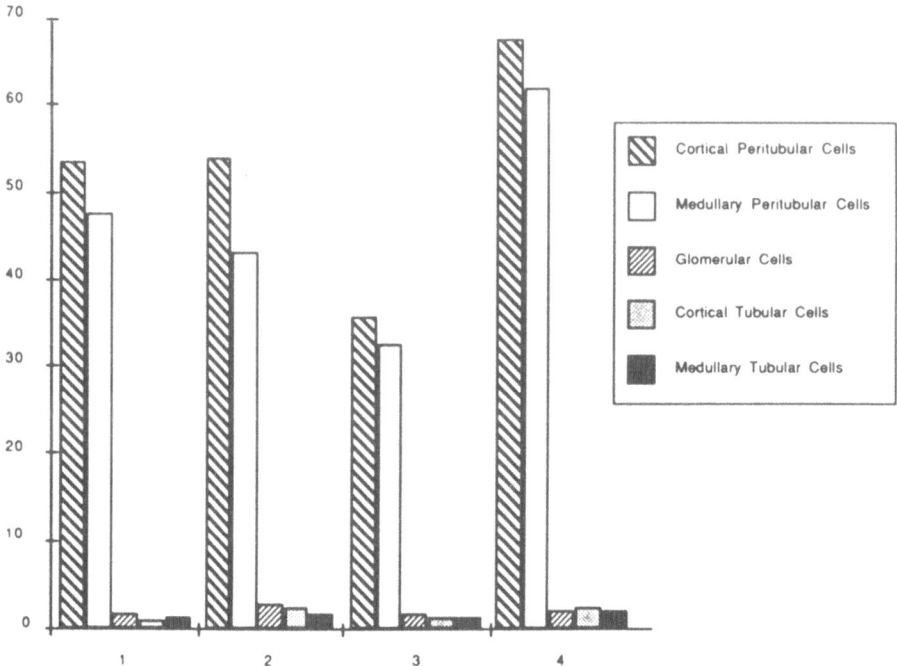

Fig. 3. Distribution of the areas of silver grains within the
different structures of mouse kidney sections hybridized with the Epo probe in four different anemic mouse
kidneys[1-4].

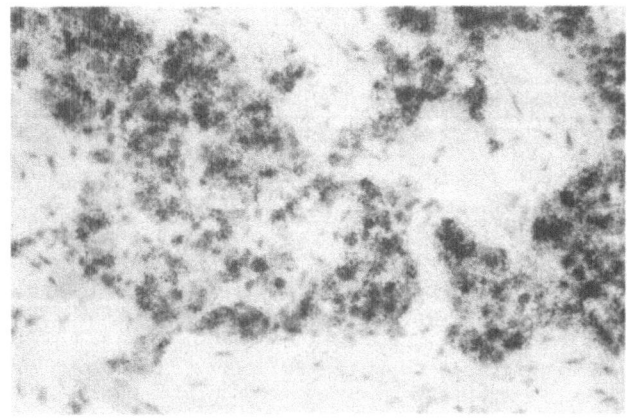

Fig. 4. In situ hybridization of tumor tissue section
from one polycythemic patient with renal
adenocarcinoma using a ^{35}S labelled monkey
Epo probe. A strong labelling is observed
in most of the tumor cells, whereas the stroma
as a background labelling (X 350).

72

results in an enhanced process of a physiological Epo mRNA expression.

In the normal kidney, there is a basal Epo production. Both renal tubular cells and endothelial cells could account for a low Epo mRNA expression, which would not be detectable by in situ hybridization. Only endothelial cells would be sensitive to the hypoxic signal, therefore in the hypoxic kidney, the endothelial cells would be positive whereas tubular cells would remain below the threshold of detection by in situ hybridization. A consequence of malignant transformation might be that the Epo gene would be permanently turned on, resulting in a high Epo expression in the tumor cells.

The second equally likelyhood explanation would be that cell transformation is often associated with "de novo" production of growth factor or hormone[19].

REFERENCES

1. Jacobson, L.O., E. Goldwasser, W. Fried, and L. Pizak. 1957. Role of the kidney in erythropoiesis. Nature (Lond) 179: 633.
2. Jelkmann, W. 1986. Renal erythropoietin: properties and production. Rev. Physiol. Biochem. Pharmacol. 104: 140.
3. Fisher, J.W., G. Taylor, and D.D. Porteous. 1965. Localization of erythropoietin in glomeruli of sheep kidney by fluorescent antibody technique. Nature 205: 611.
4. Busuttil, R.W., B.L. Roh, and J.W. Fisher. 1972. Localization of erythropoietin in the glomerulus of the hypoxic dog kidney using a fluorescent antibody technique. Acta Haematol. 47: 238.
5. Burlington, H., E.P. Cronkite, U. Reincke, and E.D. Zanjani. 1972. Erythropoietin production in cultures of goat renal glomeruli. Proc. Natl. Acad. Sci. USA 69: 3547.
6. Kurtz, A., W. Jelkmann, F. Sinowatz, and C. Bauer. 1983. Renal cell cultures as a model for study of erythropoietin production. Proc. Natl. Acad. Sci. USA 80: 4008.
7. Schuster, S.J., J.H. Wilson, A.J. Erslev, and J. Caro. 1987. Physiologic regulation and tissue localization of renal erythropoietin messenger RNA. Blood 70: 316.
8. Lacombe, C., J.L. Da Silva, P. Bruneval, J.G. Fournier, F. Wendling, N. Casadevall, J.P. Camilleri, J. Briety, B. Varet, and P. Tambourin in. 1988. Peritubular cells are the site of erythropoietin synthesis in the murine hypoxic kidney. J. Clin. Invest. 81: 620.
9. Da Silva, J.L., C. Lacombe, P. Bruneval, N. Casadevall, M. Leporrier, J.P. Camilleri, J. Bariety, P. Tambourin, and B. Varet. Tumor cells are the site of erythropoietin synthesis in human renal cancers associated with polycythemia (submitted for publication)
10. Beru, N., J. McDonald, C. Lacombe, and E. Goldwasser. 1986. Expression of the erythropoietin gene. Mol. Cell. Biol. 6: 2571.
11. Bondurant, M.C., and M.J. Koury. 1986. Anemia induces accumulation erythropoietin mRNA in the idney and liver. Mol. Cell Biol. 6: 2731.
12. McDonald, J.D., F.K. Lin, and E. Goldwasser. 1986. Cloning, sequencing, evolutionary analysis of the mouse erythropoietin gene. Mol. Cell Biol. 6: 842.
13. Salzmann, J.L., D. Bernuau, A. Poliard, C. Boussard, and G. Feldmann. 1986. An automatic method for counting silver grains in auto-historadiography: an application to in situ hybridization on rat liver cells. Acta Stereol. 5: 273.
14. Hume, D.A., and S. Gordon. 1983. Mononuclear phagocyte system of the mouse defined by immunohistochemical localization of antigen F4/80. J. Exp. Med. 157: 1704.

15. Rich, I.N., C. Vogt, and S. Pentz. 1988. Erythropoietin gene expression in vitro and in vivo detected by in situ hybridization. Blood Cells 14: 505.

16. Lin, F.K., C.H. Lin, P.H. Lai, J.K. Browne, J.C. Egrie, R. Smalling, G.M. Fox, K.K. Chen, M. Castro, and S. Suggs. 1986. Monkey erythropoietin gene: cloning expression and comparison with the human erythropoietin gene. Gene 44: 201.

17. Terreros, D.A., A. Behbehani, and F.E. Cuppage. 1986. Evidence for proximal tubular cell origin of a sarcomatoid variant of human renal cell carcinoma. Virchows Arch (A) 408: 623.

18. Koury, S.T., M.C. Bondurant, and M.J. Koury. 1988. Localization of erythropoietin synthesizing cells in murine kidneys by in situ hybridization. Blood 71: 524.

19. Sporn, M.B. and A.B. Roberts. 1985. Autocrine growth factors and cancer. Nature (Lond) 313: 745.

REGULATION OF THE ERYTHROPOIETIN GENE

Shigehiko Imagawa, Mark A. Goldberg, and H. Franklin Bunn

Howard Huges Medical Institute
Division of Hematology
Brigham and Women's Hospital, and the Department of Medicine
Harvard Medical School
Boston, MA 02115

INTRODUCTION

In man and other mammals the red cell mass is regulated by the hormone erythropoietin (Epo). This glycoprotein is produced in the fetal liver[1] and the adult kidney[2] in response to hypoxia. The hormone travels to hematopoietic tissues where it stimulates the proliferation and differentiation of erythroid progenitor cells. Recent in situ hybridization suggests that in the kidney Epo mRNA is localized in a subset of peritubular cells[3,4]. The site of Epo production in the liver remains uncertain. In addition, little is understood about the mechanism by which hypoxia leads to the increased expression of the Epo gene.

Investigation of the regulation of the Epo gene would be greatly facilitated by the availability of a cell culture system that produces significant amounts of Epo in response to hypoxia. We have recently demonstrated that the human hepatoma cell line, Hep3B, can be induced to produce large amounts of biologically active and immunologically identifiable Epo in response to hypoxia as well as cobalt, nickel, and manganese[5]. Furthermore, upon such stimulation markedly increased levels of Epo mRNA are observed. Using this system we have performed experiments which begin to elucidate the signal transduction pathway by which hypoxia stimulates Epo production. We present evidence that the oxygen sensor is a heme protein and that ligand binding to this heme protein influences Epo production and secretion. In addition, we present preliminary information on the important cis-acting regulatory regions of the Epo gene.

MATERIALS AND METHODS

Assays

The in vitro bioassay was performed by the method of Krystal[6] with several modifications[7] as described previously[8]. The radioimmunoassay (RIA) for Epo was performed using polyclonal rabbit anti-human Epo antiserum raised against human recombinant Epo. The initial RIA was performed with antiserum generously provided by Genetics Institute, Inc.[8] Subsequently a much higher titer polyclonal rabbit antiserum was produced in our laboratory using a protocol similar to that of Egrie, et al[9]. [125]I recombinant Epo was obtained from Amersham. Standards were prepared from partially

Molecular Biology of Erythropoiesis
Edited by J. L. Ascensao *et al.*
Plenum Press, New York

purified Epo (Toyobo, Epo-301) diluted in alpha MEM containing 10% fetal bovine serum. Aliquots of 0.5 ml of standard or sample were placed in 1.5 ml Eppendorf tubes. To this was added 0.1 ml of antiserum diluted 1:30,000 in phosphate buffered saline (PBS) containing 0.5% bovine serum albumin and 0.05 ml (6,000 cpm) of ^{125}I Epo diluted with the same buffer. The assay was incubated for 1 to 3 days at $4°C$ with constant shaking. 0.3 ml of Tachisorb R Immunoabsorbent (goat antibody to rabbit γ-globulin conjugated to Pansorbin Staphylococcus aureus Cells, Calbiochem) was then added to each tube and the tubes were returned to $4°C$ with constant shaking for 3 to 4 hours. The Tachisorb was pelleted by centrifugation for 30 minutes at 1500xg at $4°C$, washed once with 1.0 ml PBS, and counted in an LKB model 1282 gamma counter.

The RIA for human growth hormone was performed with a kit produced by Nichols Institute. Chloramphenicol acetyltransferase (CAT) was measured as described by Neumann et al[10].

Cell Culture

Cells were cultured in 100x20 mm tissue culture dishes (Corning No. 25020), 25 cm^2 tissue culture flasks (Corning No. 25100), or 150 cm^2 tissue culture flasks (Corning No. 25120) using MEM Alpha medium (Gibco) supplemented with penicillin (100 units/ml), streptomycin (100 µg/ml) and 10% heat inactivated ($56°C$, 30 minutes) fetal bovine serum (Gibco). Cells were maintained in a humidified 5% CO_2 incubator at $37°C$. Culture media was changed daily for 1-2 days prior to all experiments. In experiments in which cells were grown at various oxygen tensions and/or in the presence of carbon monoxide, the procedure employed was as described previously[8].

Northern Blot Analysis

Total RNA was prepared from cultured cells as described by Chirgwin, et al[11]. Northern blot analysis was performed as described previously[8]. Mouse β actin was radiolabeled and hybridized to the Northern blots in order to provide an internal control for efficiency of RNA transfer to the filters.

Electroporation

Transfection of the various plasmid DNA constructions described below into Hep3B cells was carried out by electroporation using the method of Chu and colleagues[12]. Optimal transfection was obtained when the cells were shocked with 250 volts using a capacitance of 1080µF.

RESULTS

Hypoxia Induced Epo Production in Hep3B Cells

Multiple renal and hepatic cell lines (including MDCK, LLC-PK$_1$, BHK, WRL 68, Hepa-1, CLCL, A704, CRFK, A498, ACHN, Cos-7, TCMK-1, LLC-MK$_2$, CaKi-2, IC-21, THP-1, HepG2, and Hep3B) and various primary cells (rat proximal tubular cells, rat glomerular endothelial cells, human umbilical vein endothelial cells, and HEK cells) were screened for either constitutive or hypoxia induced production of Epo. All cell lines grew well under hypoxic conditions. Only the human hepatoma cell lines, HepG2 and Hep3B, made readily measurable amounts of Epo as measured both by RIA and bioassay. Hep3B cells exhibited a much greater enhancement in Epo production in response to hypoxia and cobalt. Although there was wide variation in absolute numbers between individual experiments, the mean erythropoietin production in a 24-hour period by Hep3B cells grown in an atmosphere of 1% oxygen, 5% CO_2 was 30-fold greater than comparable cells grown in 21% O_2 and 5% CO_2 as determined by RIA[8]. Because of the marked induction of Epo production in Hep3B cells by hypoxia, this cell line was used to perform the remainder of the experiments to be described in this paper.

Effect of Metals on Epo Production

It has been known for over 40 years that cobalt increases the red cell mass in both man[13] and experimental animals[14]. Experiments with intact animals[14] as well as perfused kidneys[15] have clearly demonstrated that cobalt stimulates erythropoiesis by increasing the production of Epo. We therefore incubated Hep3B cells for 24 hours in the presence of increasing amounts of $CoCl_2$ and then analyzed the media for the presence of Epo. As shown in Table 1, cobalt greatly enhanced Epo production in Hep3B cells in a dose-dependent manner. The mechanism by which cobalt stimulates Epo production and secretion is not known. The initial suggestion[16] that cobalt acted through inhibition of cellular oxidative phosphorylation is not tenable since more potent inhibitors of cell respiration, such as cyanide, have no effect on Epo production[17].

Intrarenal injections of nickel have also been shown to induce erythrocytosis[18,19]. Hence, we incubated Hep3B cells for 24 hours in the presence of increasing concentrations of nickel chloride ($NiCl_2$) and, as summarized in Table 1, we observed a dose-dependent increase in Epo production to levels similar to those achieved with $CoCl_2$, albeit at a slightly higher molar concentration of nickel. Subsequently, we studied the effects of manganese, zinc, iron, cadmium, and tin on Epo production by Hep3B cells. Of these, only manganese induced measurable Epo levels but these levels were less than those obtained by either cobalt or nickel (Table 1). Although cobalt, nickel, and manganese are not physiologic stimuli, information on their mechanism of action could provide insights into the regulation of Epo production.

Northern Analysis

Northern blot analyses of mRNA from Hep3B cells are shown in Figure 1. Hybridization with the ^{32}P-labelled Epo cDNA probe was seen just below the 18S ribosomal RNA band, compatible with an mRNA size of slightly less than 1.8 Kb. Epo mRNA levels were greatly increased in Hep3B cells grown in a 1% oxygen and 5% CO_2 atmosphere for 24 hours compared to levels in comparable cells grown at 21% oxygen. Similarly, Hep3B cells showed a marked increase in the levels of Epo mRNA when the cells were grown in the presence of 50 to 100 μM cobalt chloride. Hybridization of the

Table 1. Effect of Various Metals on Epo Production by Hep3B Cells

Metal	Concentrations Studied (μM)	Induction of Epo	Concentration at Which Peak Effect Observed	Approximate Fold Stimulation
$Fe(NH_4)_2(SO_4)_2$	50-500	NO	--	--
$CoCl_2$	10-500	YES	50-100 μM	20-50
$NiCl_2$	10-600	YES	300 μM	20-50
$MnCl_2$	50-600	YES	100-600 μM	12-16
$ZnCl_2$	50-600	NO	--	--
$SnCl_2$	10-100	NO	--	--
$CdCl_2$	1-500	NO	--	--

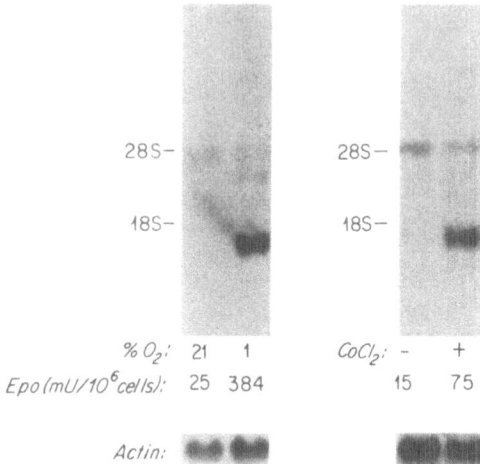

Fig. 1. Effect of hypoxia or cobalt on
Epo mRNA levels. Northern blot
analyses were performed following
growth of Hep3B cells for 24 hours
in the presence or absence of hypoxia or 50 µM CoCl$_2$. RNA was
hybridized with ^{32}P-labelled Epo
cDNA. 30 µg of total RNA was
loaded in all lanes. The film
was exposed for 6 to 7 days. The
corresponding erythropoietin protein
production by these cells is depicted
in the figure. Hybridization to
radiolabeled mouse β-actin is
shown at the bottom of the figure.
Reprinted with permission from
Goldberg, et. al.[8].

filters with the radiolabeled mouse β-actin confirmed that similar amounts
of total RNA were transferred to the filters in each lane of the paired
experiments (Figure 1).

Nature of the Oxygen Sensor

Many proteins that participate in reactions with molecular oxygen
do so via a heme moiety. Indeed, the reversible binding of oxygen to
ferroheme proteins affords a better means of "fine tuning" than the consumption of oxygen in an irreversible chemical reaction. Hemoglobin is a
classic example of such a heme protein. We propose, as depicted in Figure
2, that the oxygen sensor for the regulation of Epo production is a heme
protein which is exquisitely dependent on the oxygen tension to which
Epo-producing cells are exposed. When the oxygen tension is sufficiently
low, this heme protein is in the deoxy conformation and it triggers increased
expression of the erythropoietin gene. Conversely, when the oxygen tension
is sufficiently high, the heme protein is in its inactive oxy conformation
and does not stimulate Epo production. When either cobalt or nickel
are introduced into this environment they can substitute for ferrous
iron in newly synthesized porphyrin rings. The resulting substituted
heme protein is locked in the deoxy conformation and, like the native
deoxygenated iron heme protein, increases Epo expression.

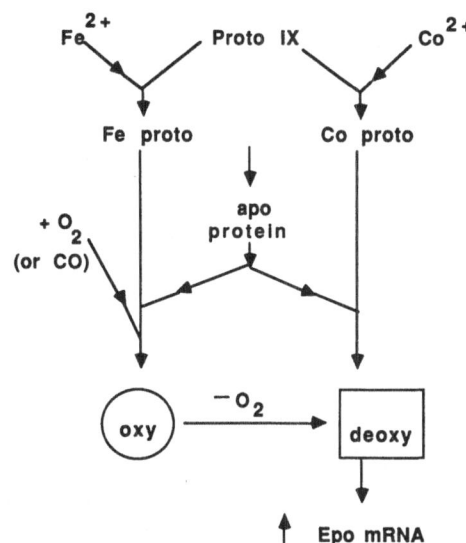

Fig. 2. Proposed model for the oxygen
sensor in Epo producing cells.
See text for detailed explanation. Cobalt's participation
in this system is also shown.
Nickel appears to work in the
same way as cobalt. Abbrevia-
tions: proto. protoporphyrin:
R. oxy conformation of the oxygen sensor: T. deoxy conformation of the oxygen sensor.

According to this model, hypoxia, cobalt, and nickel exert their
effects through a common pathway. Hence, if the Hep3B cells were maximally
stimulated by hypoxia to increase Epo expression, one would expect a
much less than additive effect when the cells are grown in the presence
of hypoxia plus either cobalt or nickel. Similarly, one would predict
no additive effect when the Hep3B cells are grown in the presence of
optimal amounts of cobalt and nickel in combination as opposed to either
stimulus alone. In fact, we observed that the effects of the combination
of hypoxia plus cobalt or hypoxia plus nickel were no different than
the effect of hypoxia alone, implying that each of these stimuli exert
their actions through the same pathway[5]. In another series of experiments
there was no additional increase in the production of Epo by Hep3B cells
grown in the presence of 50 μM $CoCl_2$ plus 300 μM $NiCl_2$ as compared to
either stimulus alone.

If the oxygen sensor is a heme protein, one would predict that carbon
monoxide would mimic oxygen and that even in the presence of hypoxia,
carbon monoxide would keep the heme protein in the "oxy" conformation
and thus prevent the hypoxia-induced increase in Epo expression. Carbon
monoxide is a highly selective ligand with regard to its metal binding
site as well as the oxidation and spin states of the metal. Reduced
heme proteins are the only known binding sites for carbon monoxide in
biological systems[20,21]. Hence, the inhibition of hypoxia-induced Epo
expression by carbon monoxide, if demonstrated, would strongly support
the hypothesis that the oxygen sensor is a heme protein. Hep3B cells
were therefore grown for 24 hours in an environment containing 1% oxygen
plus 10% carbon monoxide. Under these conditions the cells expressed

approximately 3.5 to 4-fold less Epo than identical cells grown in the presence of 1% oxygen in the absence of carbon monoxide. This is a specific effect of carbon monoxide, presumably on a heme protein, and not a non-specific toxic effect, as confirmed by the finding that Hep3B cells grown in 1% oxygen, 10% carbon monoxide, and either 100 μM $CoCl_2$ or 300 μM $NiCl_2$ make at least as much Epo as Hep 3B cells grown in 100 μM $CoCi_2$ or 300 μM $Nicl_2$ alone. Since cobalt protoporphyrin and nickel protoporphyrin cannot bind carbon monoxide, one would predict that carbon monoxide would not block the cobalt-induced or nickel-induced stimulation of Epo production. In fact, when Hep3B cells were grown in an atmosphere of 15% oxygen, 5% carbon dioxide, and either 80% carbon monoxide or 80% nitrogen, Epo expression in response to either cobalt or nickel stimulation was not statistically significantly different (data not shown), again confirming our hypothesis.

Independent confirmation that oxygen sensing depends on a heme protein can be provided by demonstrating that hypoxia-, cobalt-, and nickel-induced Epo production is inhibited by blocking heme synthesis in Hep3B cells, thereby impairing the oxygen-sensing mechanism. Hep3B cells were therefore incubated for 24 hours in media containing either desferrioxamine, a very potent iron chelator and inhibitor of heme synthesis[22], or 4.6-dioxo-heptanoic acid, an inhibitor of aminolevulinate dehydratase[23], an enzyme in the heme synthetic pathway. The cells were then grown in either a 1% oxygen environment or in the presence of 75 μM $CoCl_2$ or 300 μM $NiCl_2$ and Epo production was measured after 24 hours. 130 μM desferrioxamine caused a 5-fold reduction in hypoxia-induced Epo production which was significantly ameliorated when 300 μM ferrous ammonium sulfate was also added. Similary, 2 mM 4.6-dioxoheptanoic acid caused a 2.3-fold reduction in hypoxia-induced Epo expression as well as a 3.3-fold decrease in nickel--induced and a 4-fold decrease in cobalt-induced Epo production. The somewhat less potent effect of the 4.6-dioxoheptanoic acid compared to des-ferrioxamine may reflect the inability of the former compound to completely inhibit heme synthesis and lower cellular heme content[24].

Cis- Regulatory Elements

Comparison of the human and mouse Epo gene sequences reveals strong homology not only in the coding sequences, but also within the 5'-flanking region, the first intron, and the untranslated portion of the 3' exon[25]. The conservation of these regions suggests that they may include regulatory elements of the Epo gene. In order to investigate these candidate regula-tory elements, constructions were prepared in which portions of the Epo gene were inserted 5' and 3' to a reporter gene. The human growth hormone (GH) gene was used as a reporter gene because the product is readily secreted from cells, accumulates in the culture medium, and can be easily and accurately measured by RIA[26]. Two stretches of the human Epo gene which have highly conserved non-coding sequences have been studied:

(1) **Sequence 5A** is an 1192 base pair HindIII to Xbal fragment which extends from 378 base pairs upstream of the cap site through the first exon and the entire first intron. In order to avoid the problem of false initiation of translation from the Epo ATG start codon, this site was changed to TAG by site directed mutagenesis[27].

(2) **Sequence 3A** which consists of a 255 base pair AccI to BglII fragment that extends 67 base pairs upstream from the Epo termination codon and covers much of the 3' non-coding region that is homologous to the mouse Epo gene.

We have prepared the following constructions. In 5AOGH, 5A has been inserted upstream of OGH a PUC vector containing the human growth hormone gene but no upstream promoter. In 5AOGH3Ac, 3A has been inserted with correct orientation downstream of the GH gene. 5AOGH3Ar has 3A in its reverse orientation. Analogous constructions have been made in the vector XGH which contains the mouse metallothionein-1 promoter upstream of the GH gene.

These vectors have been transfected into the Hep3B human hepatoma cell line which, as shown above, produces Epo in a regulated fashion. We found that electroporation was the most efficient method for transfection efficiency. The production of GH was linear over a four day period. To correct for transfection efficiency, cells have been co-transfected with RSVCAT as an internal standard. At the end of either four day incubations with 50 μM $CoCl_2$ or three day incubations with 1% O_2, GH was measured in the cell media and the cell pellet was assayed for chloramphenicol acetyl transferase (CAT). The results, shown in Table II, are expressed as GH ng/CAT U. The results of transient transfections are shown in this Table. Compared with XGH, the production of GH by 5AXGH was stimulated 1.7 fold by $CoCl_2$. Furthermore, the addition of 3A 3' to the GH gene (5AXGH3A), stimulated GH production about 2.5 fold with $CoCl_2$, regardless of its orientation. Very similar results were obtained in parallel transfection experiments with constructions with OGH which lacks an artificial promoter. However the GH signal was considerably lower. In order to test the physiologic regulation of the Epo gene, the cells were incubated with hypoxia (1% O_2). Compared with XGH, the addition of the 5' element resulted in a 1.8 fold stimulation. Furthermore, with the addition of 3A 3' to the GH gene (5AXGH3A), hypoxia stimulated GH production 2.3 fold.

In contrast to the above results in Hep3B cells, we have performed parallel transfections in the COS-7 (African green monkey kidney) cell line. This cell line does not produce Epo either constitutively or following stimuli with hypoxia or cobalt and therefore is presumably lacking in the trans-acting factors necessary for regulation of the Epo gene. The results of transient transfection are shown in Table II. The addition of the EPO 5' and 3' elements failed to induce significant stimulation with either $CoCl_2$ or hypoxia.

In stable transfection experiments, Hep3B cells were co-transfected with plasmid 5AOGH3Ac and pSV-Neo (a plasmid containing the SV40 virus promoter driving the TN5 neo gene) at a ratio of 9:1. The transfected cells were grown in G418-containing medium (0.4 mg/ml) and G418-resistant colonies were isolated, grown, and further evaluated. A high proportion of these G418-resistant colonies should have also incorporated the 5AOGH3Ac plasmid. In these experiments, the Hep3B subclones were incubated for two days in either 50 μM $CoCl_2$ or in 1% O_2. In one clone, CO6, cobalt stimulated GH expression 4-fold and hypoxia increased GH production 16-fold. Furthermore, when these cells were exposed to 1% O_2 and 10% CO, the expression of GH decreased by about 3-fold compared to 1% O_2 alone. This finding is very similar to what was observed with endogenous Epo production in the CO6 cells as well as by non-selected Hep3B cells grown under similar conditions. This strongly suggests that Epo DNA sequences 5A and/or 3A contain significant cis-acting regulatory elements involved in Epo gene expression. Further experiments are currently underway to further delineate the cis-acting elements that regulate the Epo gene.

DISCUSSION

The results presented here with Hep3B cells suggest a physiologically important role for the fetal hepatocyte in the regulated production of Epo. The Hep3B cell line, as well as the HepG2 cell line, were generated

Table II. Transient Transfection Experiments.

All data in this table have been normalized to the production of human growth hormone by XGH, a construct which contains the mouse metallothionine promoter but no Epo elements. In order to correct for transfection efficiency, data are expressed as the ratio of GH to activity of CAT.

A. Hep3B cells: effect of hypoxia

	$\dfrac{[GH/CAT\ 1\%\ O_2]}{[GH/CAT\ 21\%\ O_2]}$
XGH	1.0
5AXGH	1.8 ± 0.3 *
5AXGH3Ac	2.3 ± 0.6 *
5AXGH3Ar	2.2 ± 0.4 *
	Mean ± 1 SD (n = 3)

B. Hep3B cells: effect of 50 μM cobalt chloride

	$\dfrac{[GH/CAT + CoCl_2]}{[GH/CAT - CoCl_2]}$
XGH	1.0
5AXGH	1.7 ± 0:4 *
5AXGH3Ac	2.4 ± 0.4 **
5AXGH3Ar	2.7 ± 0.5 **
	Mean ± 1 SD (n = 4)

C. COS7 cells: effect of hypoxia

	$\dfrac{[GH/CAT + 1\%\ O_2]}{[GH/CAT - 21\%\ O_2]}$
XGH	1.0
5AXGH	0.8
5AXGH3Ac	1.0
5AXGH3Ar	1.0
	(n = 1)

D. COS7 cells: effect of cobalt chloride

	$\dfrac{[GH/CAT + CoCl_2]}{[GH/CAT - CoCl_2]}$
XGH	1.0
5AXGH	0.8
5AXGH3Ac	0.8
5AXGH3Ar	1.0
	(n = 2)

* p<0.05 different from XGH
** p<0.05 different from 5AXGH

from human hepatic carcinomas[28]. Both have been shown to have histologic and biochemical characteristics of well-differentiated liver parenchymal cells[28,29]. They both have many of the biosynthetic capabilities of normal hepatocytes and have been shown to secrete 17 of the major plasma proteins into cell culture medium, including albumin and α-fetoprotein. Despite the similarities between these two cell lines and hepatocytes in intact liver, one must be careful in making analogies between gene expression in a tumor cell line and that in a normal hepatocyte. However, further evidence suggesting that this in vitro system accurately reflects the normal physiological situation in vivo is provided by the Northern blot analysis. The Epo mRNA size appears similar to that previously reported for RNA from normal human fetal liver[30] and from the kidneys of intact rodents exposed to various hypoxic stresses[31,32]. In addition, the finding that regulation occurs at the mRNA level is in agreement with previous in vivo data[31,32].

The Hep3B cell culture system has enabled us to show a direct correlation between Epo production and the level of Epo mRNA. Our results strongly support the hypothesis that the oxygen-sensing mechanism involves a specific oxygen receptor rather than a non-specific block in oxidative phosphorylation. We present evidence that the oxygen sensor is a heme protein and that ligand binding to this heme protein influences Epo production and secretion.

Preliminary experiments have allowed us to begin to delineate the cis-acting regulatory regions. It appears that some of the physiological regulation of Epo expression is conferred by conserved 5' and/or 3' regions of the Epo gene. The transient transfection experiments presented above (Table II) show only a modest (2.3-fold) stimulation in the production of the reporter gene with hypoxia or cobalt. More marked stimulation was achieved in one stably transfected clone of Hep3B cells. Further experiments are needed to determine whether more distant elements are required for the regulation of the Epo gene.

ACKNOWLEDGEMENT

This work was supported in part by NIH grant no. DK01401. The human genomic Epo plasmid and cDNA Epo plasmid were gifts from Dr. Charles Shoemaker of Genetics Institute, Inc. and the human growth hormone plasmids, pOGH and pXGH, were a gift of Dr. Richard Selden of the Department of Molecular Biology and Department of Genetics, Harvard Medical School and the Massachusetts General Hospital. Mouse β-actin cDNA was a gift from Dr. Bruce Spiegelman of the Dana-Farber Cancer Institute. Mark A. Goldberg is the recipient of a Physician Scientist Award from the National Institutes of Health and the Tomasso Family Fellowship from the Brigham and Women's Hospital.

REFERENCES

1. Zanjani, E.D., J. Poster, H. Burlington, L.I. Mann, and L.R. Wasserman. 1977. Liver as the primary site of erythropoietin production in the fetus. J. Lab. Clin. Med. 89: 640.
2. Jacobson, L.O., E. Goldwasser, W. Fried, and L. Plzak. 1957. Role of the kidney in erythropoiesis. Nature 179: 633.
3. Koury, S.T., M.C. Bondurant, and M.J. Koury. 1988. Localization of erythropoietin synthesizing cells in murine kidneys by in situ hybridization. Blood 71: 524.
4. Lacombe, C., J-L. Da Silva, P. Bruneval, J-G. Fournier, F. Wendling N. Casadevall, J-P. Camilleri, J. Bariety, B. Varet, and P. Tambourin. 1988. Peritubular cells are the site of erythropoietin synthesis in the murine hypoxic kidney. J. Clin. Invest. 81: 620.
5. Goldberg, M.A., S.P. Dunning, and H.F. Bunn. 1988. Regulation of the erythropoietin gene: Evidence that the oxygen sensor is a heme protein. Science 242: 1412.
6. Krystal, G. 1983. A simple microassay for erythropoietin based on ^{3}H-thymidine incorporation into spleen cells from phenylhydrazine treated mice. Exp. Hemtol. 11: 649.

7. Lappin, T.R.J., G.E. Elder, S.H. McKibbin, P.T. McNamee, M. McGeown, and J.M. Bridges. 1985. The effect of transferrin saturation on the estimation of erythropoietin by the mouse spleen cell microassay. Exp. Hematol. 13: 1007.

8. Goldberg, M.A., G.A. Glass, J.M. Cunningham, and H.F. Bunn. 1987. The regulated expression of erythropoietin by two human hepatoma cell lines. Proc. Natl. Acad. Sci. U.S.A. 84: 7972.

9. Egrie, J.C., P.M. Cotes, J. Lane, RE. Gaines Das, and R.C. Tam. 1987. Development of radioimmunoassays for human erythropoietin using recombinant erythropoietin as tracer and immunogen. J. Immunol. Methods 99: 235.

10. Neumann, J.R., C.A. Morency, and K.O. Russian. 1987. A novel rapid assay for chloramphenicol acetyltransferase gene expression. Bio Techniques 5: 444.

11. Chirgwin, J.M., A.E. Przybyla, R.J. MacDonald, and W.J. Rutter. 1979. Isolation of biologically active ribonucleic acid from sources enriched in ribonuclease. Biochemistry 18: 5294.

12. Chu, G., H. Hayakawa, and P. Berg. 1987. Electroporation for the efficient transfection of mammalian cells with DNA. Nucleic Acids Res. 15: 1311.

13. Berk, L., J.H. Burchenal, and W.B. Castle. 1949. Erythropoietic effect of cobalt in patients with or without anemia. N. Engl. J. Med. 240: 754.

14. Goldwasser, E., L.O. Jacobson, W. Fried, and L.F. Plzak. 1958. Studies on erythropoiesis. V. The effect of cobalt on the production of erythropoietin. Blood 13: 55.

15. Fisher, J.W., and J.W. Langston. 1968. Effects of testosterone, cobalt, and hypoxia on erythropoietin production in the isolated perfused dog kidney. Ann. New York Acad. Sci. 149: 75.

16. Yastrebov, A.P. 1966. Mechanism of cobalt action on erythropoiesis. Fed. Proc. 25: T630.

17. Necas, E., and E.B. Thorling. 1972. Unresponsiveness of erythropoietin-producing cells to cyanide. Ann. J. Physiol. 222: 1187.

18. Jasmin, G., and B. Solymoss. 1975. Polycythemia induced in rats by intrarenal injection of nickel sulfide, Ni_3S_2. Proc. Soc. Exp. Biol. Med. 148: 774.

19. Morse, E.E., T-Y. Lee, R.F. Reiss, and F.W. Sunderman. 1977. Dose-response and time-response study of erythrocytosis in rats after intrarenal injection of nickel subsulfide. Ann. Clin. Lab. Sci. 7: 17.

20. Caughey, W.S. 1970. Carbon monoxide bonding in heme proteins. Ann. New York Acad. Sci. 174: 148.

21. Coburn, R.F. 1979. Mechanisms of carbon monoxide toxicity. Prev. Medicine 8: 310.

22. Shedlofsky, S.I., P.R. Sinclair, H.L. Bonkovsky, J.F. Healey, A.T. Swim, and M. Robinson. 1987. Haem synthesis from exogenous 5-aminolaevulinate in cultured chick-embryo hepatocytes. Biochem. J. 248: 229.

23. Tschudy, D.P., R.A. Hess, and B.C. Frykholm. 1981. Inhibition of δ-aminolevulinic acid dehydrase by 4.6-dioxoheptanoic acid. J. Biol. Chem. 256: 9915.

24. Schoenfeld, N., Y. Greenblat, O. Epstein, and A. Atsmon. 1982. The effects of succinylacetone (4.6-dioxoheptanoic acid) on δ-aminolevulinate synthase activity and the content of heme in monolayers of chick embry liver cells. Biochim. Biophys. Acta 721-408.

25. Schoemaker, C.B., and L.D. Mitsock. 1986. Murine Erythropoietin: Cloning, expression, and human gene homology. Mol. Cell. Biol. 6: 849.

26. Selden, R.F., K.B. Howie, M.E. Rowe, H.M. Goodman, and D.D. Moore. 1986. Human growth hormone as a reporter gene in regulation studies employing transient gene expression. Mol. Cell. Biol. 6: 3173.

27. Kunkel, T.A., J.D. Roberts, and R.A. Zakour. 1987. Rapid and efficient site-specific mutagenesis without phenotypic selection. Methods Enzymol. 154: 367.

28. Aden, D.P., A. Fogel, S. Plotkin, I. Damjanov, and B.B. Knowles. 1979. Controlled synthesis of HBsAg in a differentiated human liver carcinoma-derived cell line. Nature 282: 615.

29. Knowles, B.B., C.C. Howe, and D.P. Aden. 1980. Human hepatocellular carcinoma cell lines secrete the major plasma proteins and hepatitis B surface antigen. Science 209: 497.

30. Jacobs, K., C. Shoemaker, R. Rudersdorf, S.D. Neill, R.J. Kaufman, A. Mufson, J. Seehra, S.S. Jones, R. Hewick, E.F. Fritsch, M. Kawakita, T. Shimizu, and T. Miyake. 1985. Isolation and characterization of genomic and cDNA clones of human erythropoietin. Nature. 313: 806.

31. Beru, N., J. McDonald, C. Lacombe, and E. Goldwasser. 1986. Expression of the erythropoietin gene. Mol. Cell. Biol. 7: 2571.

32. Bondurant, M., and M. Koury. 1986. Anemia induces accumulation of Erythropoietin mRNA in the kidney and liver. Mol. Cell. Biol. 6: 2731.

STUDIES OF THE EFFECT OF ERYTHROPOIETIN ON

HEME SYNTHESIS

Nega Beru[1] and Eugene Goldwasser[2]

[1]Department of Medicine, Joint Section of Hematology/Oncology,
University of Chicago Medical Center, Chicago, IL 60637

[2]Department of Biochemistry and Molecular Biology, The University
of Chicago, Chicago, IL 60637

INTRODUCTION

The studies described in this paper were directed at an understanding
of the regulation of heme synthesis during erythropoietin (epo)-induced
erythroid differentiation. Although considerable work has been done
on the regulation of heme synthesis in virally-transformed erythroid cell
lines which were chemically induced to differentiate along the erythroid
pathway[1-4] our goal was to study the control of heme formation in normal
bone marrow cells, the physiological site of epo action. Our earlier
findings showing that heme synthesis is required for the epo-induced
globin synthesis in marrow cells[5], made this an important problem.

The regulation of heme synthesis in the liver, where it has been
extensively studied, is at a level of δ-aminolevulinate synthase (ALA-S),
the first enzyme of the heme biosynthetic pathway. Heme exerts feedback
inhibition on activity of the enzyme and represses its synthesis[6,7]. The
regulation of heme synthesis in erythroid cells is clearly different
from that observed in hepatocytes. In prenatal liver which is predominantly
erythropoietic, heme does not repress ALA-S[8] nor is the enzyme activity
rate limiting for hemoglobin formation[9]. Incubation of marrow cells
in the presence of δ-aminolevulinate does not result in increased heme
synthesis[10]. Moreover, not only do hepatic and erythroid ALA-S exhibit
different responses to hypoxia, polycythemia, epo and porphyrogenic chemicals[11], but in contrast to the findings for δ-aminolevulinate dehydratase,
antibodies to hepatic ALA-S show no cross reactivity to the erythroid
form indicating that there may be two ALA-S genes[12].

In Friend erythroleukemic cells chemically induced to differentiate
along the erythroid pathway, there is a sequential increase in the level
of the enzymes of the heme biosynthetic pathway taking place over a period
of four days. The synthesis of heme however does not follow induction
of ALA-S but that of ferrochelatase suggesting, that in these cells,
the terminal step of heme synthesis may be the rate limiting step[2-4].
Studies of the regulation of heme biosynthesis during epo-induced erythroid
differentiation have also been conducted. While some investigators found
that epo caused a sequential induction of ALA-S, δ-aminolevulinate dehydratase and ferrochelatase[13], others found that in embryonic mouse liver

Molecular Biology of Erythropoiesis
Edited by J. L. Ascensao *et al.*
Plenum Press, New York

cells which are responsive to epo, epo induced the synthesis of hemoglobin but no increase in activity for the three enzymes was found[14]. The activity of porphobilinogen deaminase, the third enzyme of the heme biosynthetic pathway, was found to be increased in response to epo in human marrow cells in culture[15]. Our studies, done in rat marrow cells in vitro and in vivo show that porphobilinogen deaminase is the only enzyme among those studied in the pathway which was induced by epo and that this may be the key enzyme in the regulation of epo-induced heme synthesis[16].

The Effect of Erythropoietin on Hemoglobin Synthesis in Suppressed Marrow Cells

Study of the effect of epo on hemoglobin synthesis in suppressed marrow cells was necessitated by the fact that normal marrow contains erythroid cells in various stages of development. Thus in order to measure the effect of epo on hemoglobin synthesis in the progenitor cell, we have used marrow from suppressed rats which is depleted of mature and maturing erythroid cells[16]. Epo has a significant effect (35% increase) on hemoglobin synthesis as early as 5 hrs after its addition to suppressed marrow cells in cultures (Fig. 1). By three days in culture there was a 95 fold increase in hemoglobin synthesis in suppressed marrow cells due to epo.

In addition to measuring the epo effect on hemoglobin synthesis by measuring ^{59}Fe incorporation into heme (almost all of the heme extractable from marrow cells is derived from hemoglobin[17,18]) one can also measure the amount of newly synthesized globin. The results of such an experiment are depicted in Fig. 2 which is an fluorogram of an SDS-polyacrylamide

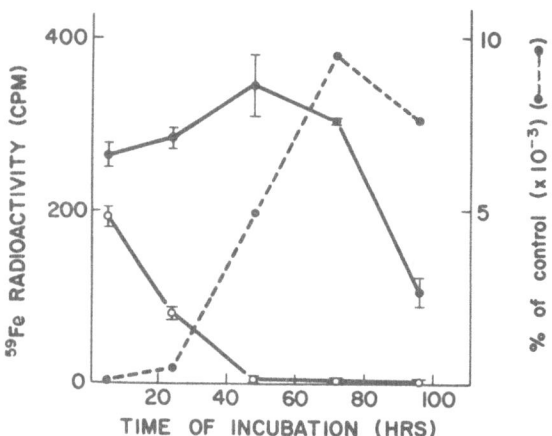

Fig. 1. The effect of epo on hemoglobin synthesis in suppressed rat marrow cells. Two-tenth ml of suppressed rat marrow cells at 20x10^6 cells/ml were incubated with (●——●) or without (o———o) 0.1U/ml epo for up to 96 hrs. At the indicated times, hemoglobin synthesis was determined by measuring ^{59}Fe incorporation into heme[16]. Each point is the mean of six replicate and the vertical bars indicate the standard deviations. ●---●, per cent of control.

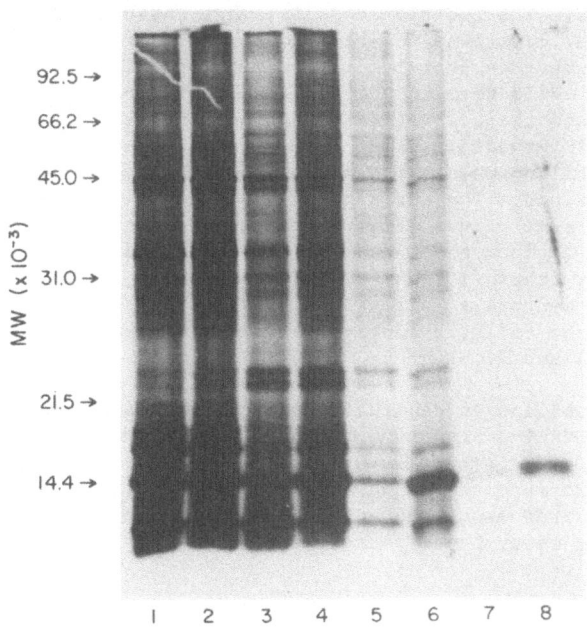

Fig. 2. The effect of epo on hemoglobin synthesis
         in suppressed rat marrow cells. One ml
         of suppressed rat marrow cell suspensions
         at 20×10^6 cells/ml were incubated with and
         without epo (0.1U/ml). At 0, 19, 43, or
         67 hrs, 20 µCi of ^{3}H-leucine (52.2 Ci/mmol
         was added and the cultures terminated 5 hrs.
         later). The relative rates of hemoglobin
         synthesis were estimated by measuring the
         relative rates of globin synthesis after the
         cell lysates were subjected to SDS poly-
         acrylamide gel electrophoresis and the globin
         bands (migrating near the 14.4 kilodalton
         marker) were visualized by fluorography and
         the intensity of the globin bands determined
         by densitometer scanning. Lanes 1 and 2,
         0-5 hrs; lanes 3 and 4, 19-24 hrs; lanes 5
         and 6, 43-48 hrs; lanes 7 and 8, 67-72 hrs.
         Lanes 1, 3, 5 and 7 are controls; while
         lanes 2, 4, 6 and 8 correspond to epo-
         treated cells.

gel electrophoresis of total cellular proteins from suppressed marrow
cells incubated with or without epo for up to four days in culture. Epo
had a significant stimulatory effect at all four time points studied.
The effect is most noticeable at 48 and 72 hrs of culture when epo caused
a four and seven fold increase respectively in globin synthesis with
the rest of the cellular proteins being relatively unaffected.

The Effect of Erythropoietin on the Activities of the Enzymes of the Heme Biosynthetic Pathway

There has to be increased rate of heme synthesis to account for
the increased rate of hemoglobin synthesis in suppressed marrow cells

in response to epo. The question thus is, what changes take place in the heme biosynthetic pathway to accommodate the increased demand for heme during epo-induced erythroid differentiation? To answer this question suppressed marrow cells were incubated with or without 0.1U/ml of epo for up to 96 hrs and the activities of various enzymes of the pathway assayed at given intervals of time. Our results, reported elsewhere[16] show that porphorbilinogen deaminase is the only enzyme induced by epo by as much as 3.5 fold. The activity of this enzyme was increased by epo to above zero-time values (Fig. 3) and in a dose-dependent fashion. Experiments using actinomycin D and cycloheximide indicate that induction is at the level of transcription. Unlike the murine erythroleukemic system where heme synthesis follows the induction of ferrochelatase, in suppressed marrow cells it appears to follow that of porphobilinogen deaminase (Figs. 1 and 3).

Unlike porphobilinogen deaminase, the activities of the other enzymes assayed, ALA-S, DOVA (α, δ-dioxovalerate) transaminase (a possible source of δ-aminolevulinate[19]) and ferrochelatase did not show any increase in activity. There appeared to be an increase in δ-aminolevulinate dehydratase activity as a result of epo action, but this increase is probably an artefact of the assay conditions[16].

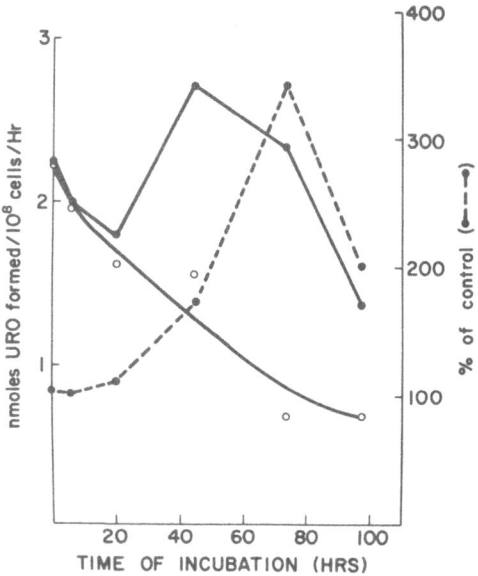

Fig. 3.   The effect of epo on phorpho-
bilinogen deaminase activity
in suppressed rat marrow cells.
Marrow cell suspensions (5ml) at
20×10^6 cells/ml were incubated
with ($\bullet$——$\bullet$) or without (o——o)
0.1U/ml epo. At the indicated
times, the cultures were terminated with prophobilinogen deaminase activity determined as nmole
uropurphyrin (URO) formed per 10^8
cells per hr[16]. Each value is the
mean of four replicate determina-
tions, $\bullet$---$\bullet$, per cent of control.

Hemoglobin Synthesis and Porphobilinogen Deaminase Activity in Marrow
Cells Exposed to Varying Amounts of Epo in Vivo

   Suppressed, normal and hyperplastic marrow cells were obtained as
described[16] and hemoglobin synthesis and porphobilinogen deaminase activity
determined. Hyperplastic marrow cells, exposed to elevated epo levels
synthesized, on a per cell basis, much more hemoglobin than either normal
and suppressed marrow cells the latter of which are exposed to depressed
levels of epo. The hyperplastic marrow synthesized 12 fold more hemoglobin
than did normal and 45 fold more than suppressed cells (Table 1). Similar
results are obtained when globin synthesis rates are examined (Fig. 4).

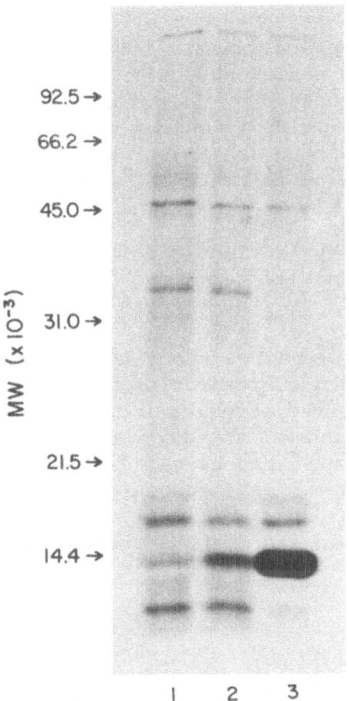

Fig. 4. Comparison of the rates of
         hemoglobin synthesis in
         suppressed, normal and hyper-
         plastic marrow cells. Sup-
         pressed, normal and hyperplastic
         marrow cells were obtained as
         described[16] and following sus-
         pension in complete medium at
         20×10^6 cells/ml were incubated
         with 20 µci of ^{3}H-leucine (52.2
         Ci/mmol) per ml of cell culture
         for 5 hrs. Total cell lysates
         equivalent to 10^6 were electro-
         phoresed on SDS polyacrylamide
         gels and hemoglobin synthesis
         measured by the rates of globin syn-
         thesis following visualization of the
         14.4 kilodalton marker) by fluoro-
         graphy. Lane 1, suppressed; lane 2,
         normal; lane 3, hyperplastic marrow.

Table 1. Hemoglobin Synthesis and Porphobilinogen Deaminase Activity
in Supressed, Normal and Hyperplastic Marrow Cells

Type of MARROW	Hemoglobin Synthesis %($Fe Heme (CPM)	Porphobilinogen Deaminase Activity nmol URO per 10^8 cells Per Hour
Suppressed	25 ± 4	3.0 ± 0.6
Normal	91 ± 10	8.2 ± 0.7
Hyperplastic	1120 ± 80	19.8 ± 2.8

Bone marrow cells from the three groups of rats (hematocrits of 72 ± 2,
36 ± 3 and 50 ± 3 for supressed, anemic and normal rats respectively)
were suspended in complete medium at $20x10^6$ cells per ml and the rate
of hemoglobin synthesis measured by determining ^{59}Fe incorporation into
heme as described[16]. Each value is a mean of six replicate determinations.
The activity of porphobilinogen deaminase was also determined in each
type of marrow and each value is the mean of four replicate determinations.

Hyperplastic marrow cells synthesized globin at a level which was 24
fold greater than that of normal cells and 48 fold greater than that
of normal cells and 48 fold greater than that of suppressed marrow cells.

When porphobilinogen deaminase activity was assayed in the three
types of marrow, there was a corresponding difference in activity. Hyperplastic marrow cells exhibited porphobilinogen deaminase activity which was
2.5 and 6.6-fold more than normal and suppressed marrow cells respectively.
Hyperplastic marrow cells also had elevated levels of δ-aminolevulinate
dehydratase but again this may be a result of the assay conditions[16].
On the other hand there was no significant difference in the activities
of the mitochondrial enzymes ALA-S, DOVA transaminase and ferrochelatase
between hyperplastic, normal and suppressed marrow cells.

CONCLUSIONS

Porphobilinogen deaminase may be the primary enzyme involved in
the control of heme synthesis during epo-induced erythroid differentiation.
Porphobilinogen deaminase is induced by erythropoietin in a dose dependent
manner and our data indicate that this induction is at the level of transcription. Moreover the time course of induction of this enzyme proceeded
in parallel fashion with the increase in hemoglobin synthesis in suppressed
marrow cells. Our in vitro studies were supported by results obtained
from our in vivo studies. In the marrow of anemic rats where a large
number of cells are in the process of erythroid differentiation due to
the increased serum epo levels, porphobilinogen deaminase activity was
much higher than in the marrow of normal rats. Conversely, in polycythemic
rats where very little erythroid differentiation is taking place in the
marrow due to lowered serum epo levels, porphobilinogen deaminase activity
is lower than in normal marrow cells. Some investigators have shown
that under erythropoietic stress caused by hypoxia, the activities of
porphobilinogen deaminase and porphobilinogen oxygenase, which removes
porphobilinogen from the heme biosynthetic pathway by oxidation to oxopyrro-
linones, are inversely correlated[20] the former reaching maximum activity
between 4-8 days of hypoxia while the later showed minimum activity at
these times.

Recently the porphobilinogen deaminase gene was cloned[21,22] and shown to be expressed both in a tissue specific and housekeeping manner. A single porphobilinogen deaminase gene is transcribed from two different promoters giving rise to two mRNA's differing only at their 5' ends and giving rise to two isoforms of the enzyme. One promoter is active in all tissues while the second is active only in erythroid cells. It is conceivable that epo, by mechanisms that still need to be elucidated, induces transcription only from the "erythroid" promoter.

REFERENCES

1. Ebert, P.S., and Y. Ikawa. 1974. Induction of δ-aminolevulinic acid synthetase during erythroid differentiation of cultured leukemia cells. Proc. Soc. Exp. Biol. Med. 146: 601.
2. Sassa, S., S. Granick, C. Chang, and A. Kappas. 1975. Induction of enzymes of the heme biosynthetic pathway in Friend leukemia cells in culture, In: Nakao, K., Fisher, I.W., Takaku, F. (eds): Erythropoiesis. Tokyo, University of Tokyo Press. p 383.
3. Sassa, S. 1976. Sequential induction of heme pathway enzymes during erythroid differentiation of mouse Friend leukemia virus-infected cells. J. Exp. Med. 143: 305.
4. Rutherford, T., G.G. Thompson, and M.R. Moore. 1979. Heme bio-synthesis in Friend erythroleukemia cells: Control by ferrochel-atase. Proc. Natl. Acad. Sci. USA 76: 833.
5. Beru, N., K. Sahr, and E. Goldwasser. 1983. Inhibition of heme synthesis in bone marrow cells by succinylacetone: Effect on globin synthesis. J. Cell Bioch. 21: 93.
6. Sassa, S., and A. Kappas. 1981. Genetic, metabolic and biochemical aspects of the porphyrias, In: Harris H., Hirshhorn, K.H.(eds): Advances in human genetics. New York, Plenum Press. p 121.
7. Granick, S., and S. Sassa. 1971. δ-Aminolevulinic acid synthetase and the control of heme and chlorophyll synthesis, In Vogel H.J. (ed): Metabolic Regulation, Vol. 5. New York, Academic Press, p 77.
8. Woods, J.S., and R.L. Dixon. 1972. Studies of the perinatal differences in the activity of hepatic δ-aminolevulinic acid synthetase. Biochem. Pharmacol. 21: 1735.
9. Freshney, R.I., and J. Paul. 1971. The activities of three enzymes of haem synthesis during hepatic erythropoiesis in the mouse embryo. J. Embryol. Exp. Morphol. 26: 313.
10. Hrinda, M.E., and E. Goldwasser, 1968. Studies on the control of hemoglobin formation in marrow cells. Ann. N.Y. Acad. Sci. 149: 412.
11. Wada, O., S. Sassa, F. Takaku, Y. Yano, G. Urata, and K. Nakao. 1967. Different responses of the hepatic and erythropoietic δ-aminolevulinic acid synthetase of mice. Biochim. Biophys. Acta. 148: 585.
12. Yamamoto, M., H. Fujita, N. Watanabe, N. Hayashi, and G. Kikuchi. 1986. An immunochemical study of α-aminolevulinate synthese and δ-aminolevulinate dehydratase in liver and erythroid cells of rat. Arch. Bioch. Biophys. 245: 76.
13. Nakao, K., S. Sassa, O. Wada, and F. Takaku. 1968. Enzymatic studies on erythroid differentiation and proliferation. Ann. N.Y. Acad. Sci. 149: 224.
14. Freshney, R.I., J. Paul, and D. Conkie. 1972. Effect of erythro-poietin on haemoglobin synthesis and haem synthesizing enzymes of mouse foetal liver cells in culture. J. Embryol Exp. Morphol. 27: 525.
15. Sassa, S., and A. Urabe. 1979. Uroporphyrinogen I synthase in-duction in normal human bone marrow cultures: An early and quantitative response of erythroid differentiation. Proc. Natl. Acad. Sci. USA 76: 5321.

16. Beru, N., and E. Goldwasser. 1985. The regulation of heme biosyn-
    thesis during erythropoietin-induced erythroid differentiation.
    J. Biol. Chem. 260: 9251.
17. Bedard, D.L., and E. Goldwasser. 1976. On the mechanism of erythro-
    poietin-induced differentiation XV. Induced transcription
    restricted by cytosine arabinoside. Exp. Cell. Res. 102: 376.
18. Gallien-Lartigue O., and E. Goldwasser. 1964. Hemoglobin synthesis
    in marrow cell culture; the effect of rat plasma on rat cells.
    Science 145: 277.
19. Varticovski, L., J.P. Kushner, and B.F. Burnham. 1980. Biosynthesis
    of porphyrin precursors: Purification and characterization
    of mammalian L-alanine: α, δ-dioxovaleric acid aminotransferase.
    J. Biol. Chem. 255: 3742.
20. Frydman, R.B., M.L. Tomaro, A. Sburlati, and A. Gutnisky. 1986.
    The regulation of porphobilinogen oxygenase and porphobilinogen
    deaminase activities in rat bone marrow under conditions of
    erythropoietic stress. Biochim Biophys Acta. 870: 520.
21. Grandchamp, B., H. De Verneuil, C. Beaumont, S. Chretien, O. Walter,
    and Y. Nordmann. 1987. Tissue specific expression of porphobilin-
    ogen deaminase: Two isoenzymes from a single gene. Eur. J.
    Biochem. 162: 105.
22. Chretien, S., A. Dubart, D. Beaupain, N. Raich, B. Grandchamp, J.
    Rosa, M. Goossens, and R. Paul-Henri. 1988. Alternative trans-
    cription and splicing of the human porphobilinogen deaminase gene
    result either in tissue-specific or in housekeeping expression.
    Proc. Natl. Acad. Sci. USA. 85: 6.

EFFECTS OF HEMIN ON ERYTHROPOIESIS

Blanche P. Alter, J. Matthew Schofield,
Liya He, and Rona S. Weinberg

The Polly Annenberg Levee Hematology Center
Departments of Medicine and Pediatrics
Mount Sinai School of Medicine
New York, NY 10029

INTRODUCTION

Hemin, an integral part of hemoglobin, also has activity as an erythroid growth factor in vitro[1-5]. Incorporation of hemin into murine or human erythroid cultures leads to an increase in the number of CFU-E or BFU-E-derived colonies. Hemin has effects in nonerythroid systems as well, stimulating growth of normal and leukemic myeloid colonies[6], acting as a T cell mitogen[7], differentiating 3T3 cells into adipocytes[8], and neuroblastoma cells into neurites[9], and promoting growth of fibroblasts[10].

Pertaining to hemoglobin synthesis itself, hemin induces the production of embryonic and fetal hemoglobin in human leukemic cell line K562[11], and minor adult hemoglobin in mouse erythroleukemic cells[12]. It also accelerates maturation of anemic rabbit erythroblasts[13], decreases DNA synthesis and induces globin RNA synthesis in immature erythroblasts[14], is required for in vitro globin synthesis[15], and for initiation of globin synthesis by binding of mRNA to ribosomes[16,17].

Patients with hemolytic anemias such as sickle cell disease or thalassemia have increased levels of plasma hemin in vivo[18]. To determine whether this might be related to their in vivo erythropoiesis and fetal hemoglobin (Hb F, $\alpha_2\gamma_2$) levels, we examined the role of hemin on these in vitro. We reported previously our preliminary studies in which we found that hemin added to in vitro erythroid cultures increased the number of CFU-E and BFU-E-derived colonies using normal or sickle marrow or blood as the starting material[19]. The effect of hemin on the relative level of γ globin chain synthesis was inconsistent.

We report here our studies to examine the temporal effect of hemin addition, and to investigate the mechanism of the hemin effect on colony growth.

METHODS

Erythroid cultures were established in methyl cellulose with peripheral blood mononuclear cells obtained from donors with Hb AA, SS, SC, and S/β thalassemia, using methods described elsewhere[20]. In some experiments, mononuclear cells (MNC) were depleted of adherent cells by 2 hr adherence

to plastic dishes[21], and the nonadherent cells (NAC) cultured. The erythropoietins (Ep) used in these experiments were human urinary Ep, either
CAT-1, 1140 u/mg, kindly provided by the National Institutes of Health
Heart, Lung, and Blood Division, or from Toyobo New York, 57 u/mg. Hemin
was prepared freshly as described[19], and either included in the cultures
on day 0, or added to each 0.3 ml well in 30 µl volumes at specified
later times. Final concentrations ranged from 50 to 800 µM in some experiments. 100 µM hemin was used for many of the Hb AA studies, and 100
or 200 µM in the sickle experiments.

Globin chain synthesis was evaluated by addition of 100 µC of ^{3}H-leucine
for 20 hrs, followed by analysis using Triton acid urea polyacrylamide
gel electrophoresis and fluorography as described[20,22]. The relative
proportion of γ globin synthesis = 100 x γ/(γ + β). Colonies/100,000
cells plated and globin synthesis data are presented as the mean ± 1
standard deviation of at least triplicate wells, and the data compared
using Student's t test[23]. Differences are considered significant if
p <0.05.

RESULTS AND DISCUSSION

Hemin Dose Response Data   Analysis of erythroid colony growth on
day 13 in the presence of 50 to 800 µM hemin confirmed our previous observation that 100 to 200 µM hemin was the most effective concentration
in cultures of blood from Hb AA donors[19]. A large series using only
these concentrations is shown in Fig. 1. 100 µM hemin led to 1 to 4-fold
increases in colony number (mean 2-fold), and 200 µM hemin led to 1.2
to 8-fold increases (mean 3.5-fold). The magnitude of the increase was
not related to the number of colonies without added hemin.

The proportion of γ globin synthesis was significantly increased
with hemin in only two experiments, significantly decreased in two, and
unchanged in four others. These results might be interpreted to indicate
that hemin increases Hb F synthesis in the cultures of a few normal individuals.

This interpretation must be made with caution, however. We had
the opportunity to study one normal individual six times over a three
year period (Fig. 2). The number of colonies on day 13 consistently
increased 1.3 to 3.2-fold (mean 1.8) in the presence of 100 µM hemin.
In two experiments, the % γ globin synthesis appeared to increase, but
these results were not significant because of wide standard deviations

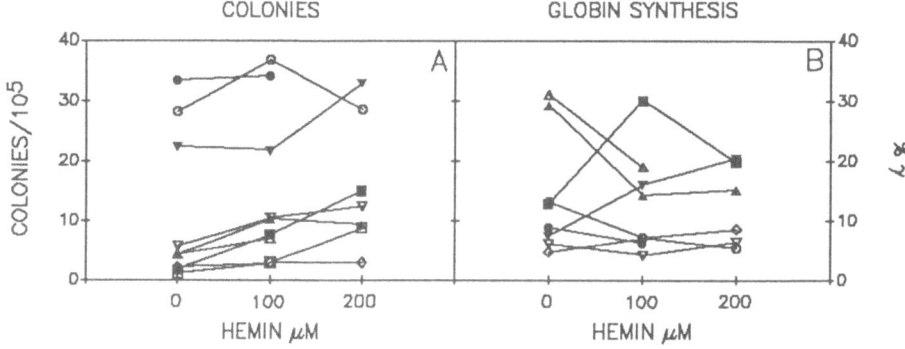

Fig. 1.   Colony growth (A) and γ globin synthesis (B) in Hb AA cultures
at 0, 100, and 200 µM hemin. Each symbol represents a different
donor.

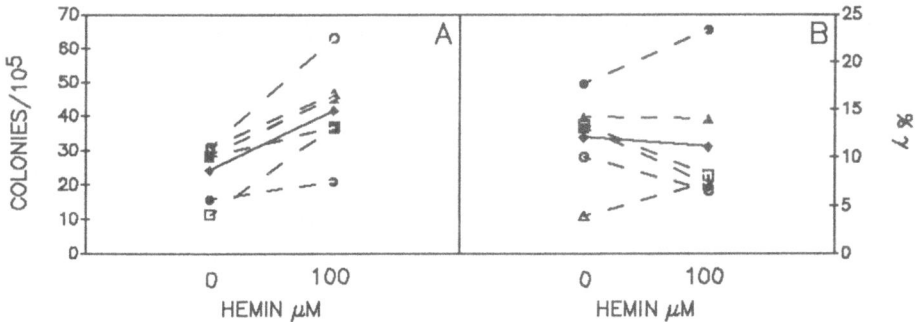

Fig. 2. Colony growth (A) and γ globin synthesis (B) in six studies
of one Hb AA donor at 0 and 100 and 100 μM. -- each study.——
mean for all studies.

(not shown). The % γ synthesis was significantly decreased in one study,
and was unchanged in the other three. Reproducibility of data within
experiments is poor, since we use 0.3 ml wells which may have as few
as 5-10 colonies. This differences in globin synthesis without and with
hemin must be large in order to be significant.

We performed similar studies using blood from several donors with
sickle syndromes (Fig. 3). 100 or 200 μM hemin increased colony numbers
1.1 to 3.7-fold (mean 1.8-fold) and 0.8 to 3.5-fold (mean 2-fold) respectively. The magnitude of the effect of hemin on sickle cultures was less
than on normal cultures, but the concentration range for increased colony
growth was broader in the sickle than in the normal studies[19] (and data
not shown). Relative γ globin chain synthesis was significantly increased
in one sickle study, and not significantly affected in all the others.

Time Course   Fig. 4 shows the results of time course studies of
colony numbers and γ globin synthesis without and with hemin in Hb AA
cultures. The peak colony numbers were generally seen on days 13-16.
The addition of 100 μM hemin on day 0 increased the numbers of colonies
at all times, without affecting the time of peak growth. During the
interval of our studies, there was a temporal decline in the % γ globin
synthesis in only one control study. With hemin included in the cultures,
there was also a decrease in the %γ globin synthesis over time in the
same one study. Thus hemin did not prevent a temporal decrease in the

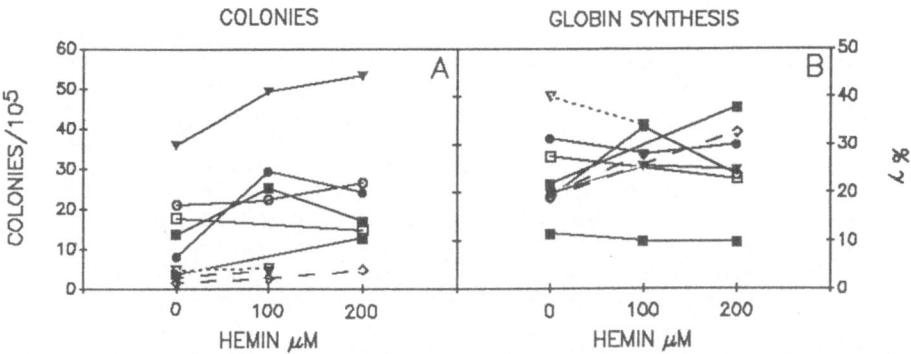

Fig. 3. Colony growth (A) and γ globin synthesis (B) in Hb S cultures
at 0, 100, and 200 μM hemin. —— Hb SS, -- Hb SC, and ···
Hb S/β thalassemia.

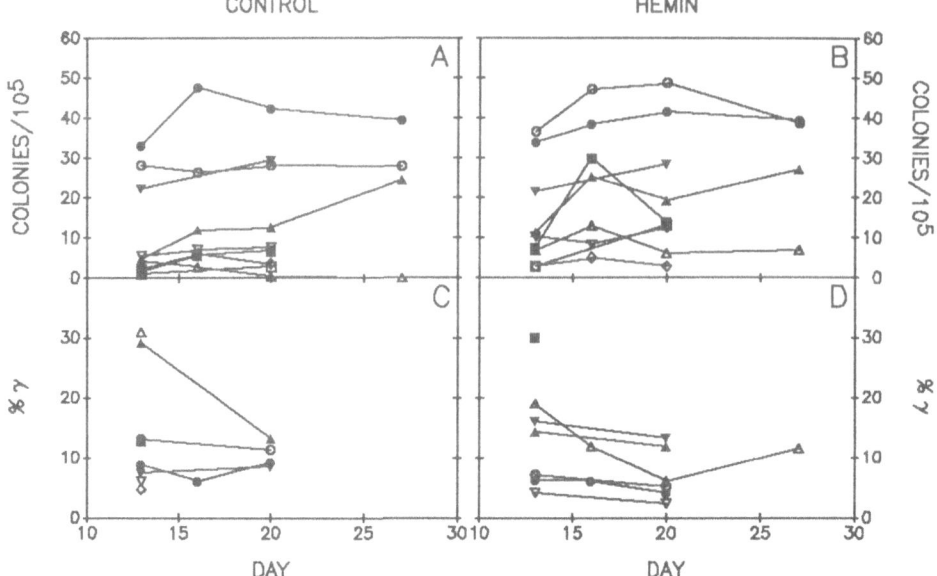

Fig. 4.   Time course in Hb AA cultures. (A) Colony growth, no hemin.
(B) Colony growth with 100 μM hemin. (C) % γ globin synthesis,
no hemin. (D) % γ globin synthesis with 100 μM hemin.

proportional synthesis of γ globin when it occurred in the absence of
hemin.

     In the sickle experiments shown in Fig. 5 the peak number of colonies
was clearly seen on a later day than in the Hb AA cultures, occurring
on days 16-20 rather than 13-16. The inclusion of hemin on day 0 increased
the numbers of colonies at all points, without further altering the timing.
The later peak growth might be due to the presence in sickle blood of
BFU-E that are less mature than those in normal blood, and thus take
longer to complete colony formation in culture. Consistent with this
suggestion is the observation that the % γ globin synthesis was higher
in the sickle colonies (mean 23%, range 11-40%) than in those derived
from normal controls (mean 14%, range 5-31%). In addition, most of the
sickle cultures did show a temporal decline in γ globin synthesis, which
was completed later in the sickle than in the normal cultures. Hemin
led to an increase in γ globin synthesis on day 16 in two of eight cultures,
but did not otherwise interfere with the temporal decline.

     Delayed Addition of Hemin   In several experiments, hemin was added
to the cultures at various times from day 0 to day 17, and colonies examined
on days 13 and 20 (Fig. 6). In the normal (Hb AA) cultures, the addition
of hemin at almost any time led to increased numbers of colonies, perhaps
due to enhanced hemoglobinization and hence easier recognition of mature
colonies. In two of the four experiments, relative γ globin synthesis
was increased on day 13-14, when hemin was added on day 7-10. However,
these results were not statistically significant. Similarly, in one
study the addition of hemin on day 16 led to increased γ globin synthesis
on day 20-21, but again this increase was not to a level that was significant.

     Similar studies of the effect of the delayed addition of hemin to
cultures from Hb SS, SC, and S/β thalassemia are shown in Fig. 7. Colony
numbers were in general much higher on day 20 than on day 13, as noted
above, and were even higher when hemin was added within the first 13

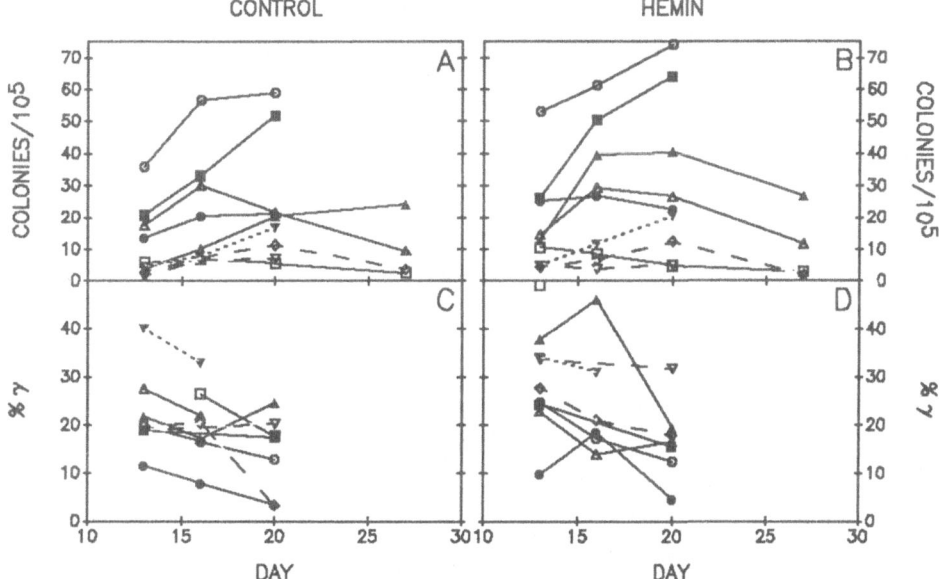

Fig. 5.   Time course in Hb A cultures. Panels as in Fig. 4, symbols as in Fig. 3.

days. This suggests that hemin might be stimulating growth of colonies from immature progenitor cells. The proportion of γ globin synthesis appeared to be increased on day 13-14 when hemin was added on day 10 in one study and day 13 in another, and on day 20-21 when hemin was added on day 7-10 in one study.

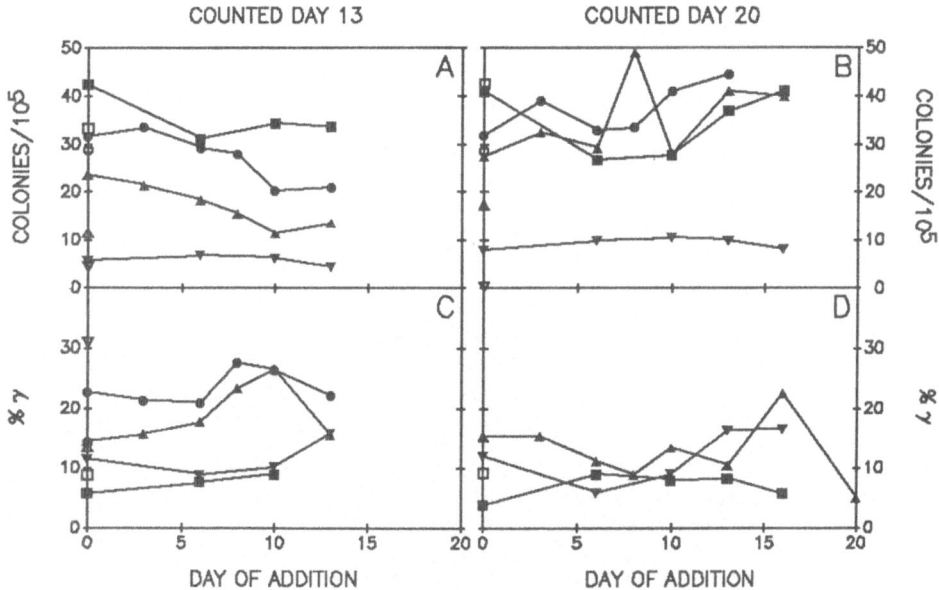

Fig. 6.   Delayed addition of 100 μM hemin to Hb AA cultures. (A) Colonies counted on day 13. (B) Colonies counted on day 20. (C) % γ globin synthesis on day 14. (D) % γ globin synthesis on day 21.

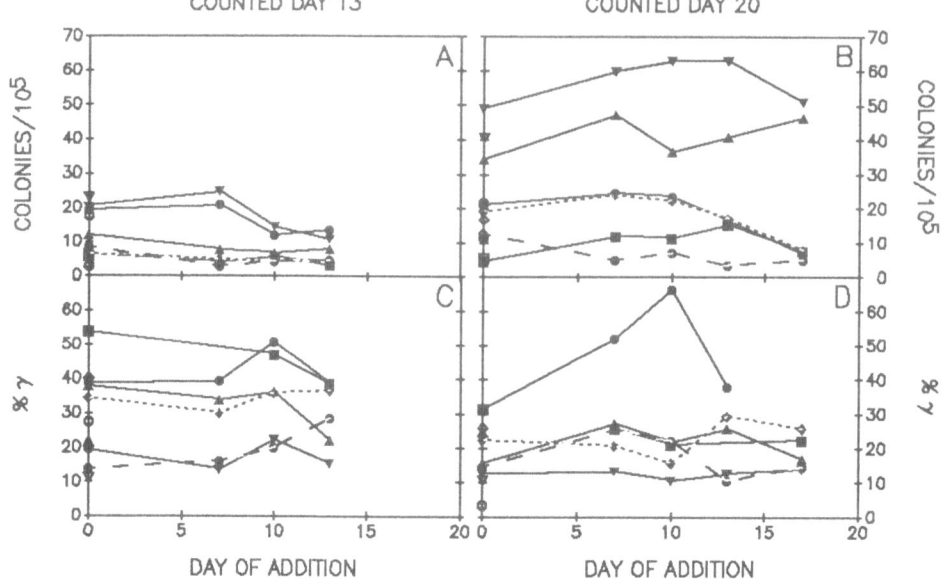

Fig. 7. Delayed addition of 200 μM hemin to Hb S cultures. Panels
as in Fig. 6, symbols as in Fig. 3.

     Colony number was thus increased when hemin was added at almost
any time. γ globin synthesis, however, was only sometimes increased
when hemin was present towards the end of the culture period. This period
of time suggests that hemin might sometimes alter the globin synthesis
pattern of the erythroblasts within the colonies in some cultures.

     Mechanism of Hemin Effect   Since peroxides are produced by cultured
cells, and hemin has a peroxidative activity, we examined the effect
of an antioxidant enzyme, catalase, on erythroid colony growth from a
normal donor[24]. There were 29 ± 6 colonies per 100,000 cells plated
without any additive, and 20-26 colonies with 0.1-25 units of catalase
per ml. In the same experiment, 100 μM hemin led to 37 ± 7 colonies
100,000 cells. Thus the concentrations of catalase used could not duplicate
the hemin effect, which may therefore not be by virtue of peroxidases.

     Another question was whether accessory cells are required for the
hemin effect. To begin to address this, we removed adherent cells from
blood mononuclear cells, and cultured the MNC and the NAC populations.
In two studies we found that the dose response effect of hemin on increasing
cell growth was maintained in the NACs. In the study from a Hb AA donor,
colony numbers per 100,000 MNC rose from 2 without to 21 with hemin,
and in the NAC from 8 to 39 per 100,000 cells plated. Similarly, in
the Hb SS experiment, MNC colony numbers were 21 without and 28 with
hemin, and the results in the sickle NACs were 20 and 41 respectively.
These results indicate that adherent cells are not involved in the hemin-
induced stimulation of erythroid colony growth.

     It was previously noted that removal of adherent cells led to a
decrease in the relative proportion of Hb F synthesis in BFU-E-derived
colonies[25]. To determine whether hemin would restore the Hb F level,
we examined globin synthesis in cultures of Hb AA MNCs and NACs, without
and with added hemin. In this experiment, γ synthesis was 18% in the
MNC and decreased to 7% in the NAC without hemin. With hemin included
from day 0, γ synthesis was 24% in the MNC and 12% in the NAC. Thus

hemin increased γ synthesis by approximately 5-6% in both the MNC and NAC cultures. Hemin and adherent cells may act independently to stimulate relative γ globin synthesis.

SUMMARY AND CONCLUSIONS

It is clear that in vitro hemin increases the number of blood BFU-E derived colonies from normal donors. This occurs with sickle donors as well, despite the increased levels of hemin in vivo in these patients. The effect of hemin on relative γ globin synthesis is inconsistent, however. In a few cases, delayed addition of hemin led to increased γ globin synthesis. In time course studies of cultures from normal donors, hemin added on day 0 shifted the day of peak colony number from 13-14 to 16-20 days. The temporal decline in γ globin synthesis was not altered. In cultures from sickle donors we found that the time for maximal colony number was later than in normals, occurring at 16-20 days even without hemin, and was not further delayed by hemin. The relative proportion of γ globin synthesis was higher on day 14 in the sickle than the normal cultures, and the temporal decline was somewhat slowed in the sickle cultures by hemin. The elevated γ synthesis and the later time for peak colony growth in the sickle cultures suggest that the erythroid progenitors in the blood of the sickle patients are less mature than those from normal individuals.

There are several possible explanations for the detection of increased numbers of colonies in cultures containing hemin. Hemin may delay the final maturation of erythroblasts within erythroid colonies, thus shifting the time of maximal growth. It may also increase the extent of final maturation, leading to more complete hemoglobinization of the erythroblasts within the colonies, and thus increasing the number of colonies that are eventually recognized as erythroid. The colonies appear to be larger in the presence of hemin, again possibly related to more complete hemoglobinization. Hemin might also lead to differentiation and maturation from immature BFU-E, which would not have developed into recognizable colonies without hemin.

ACKNOWLEDGEMENTS

This work was supported in part by National Institutes of Health grants HL26132 and National Foundation March of Dimes grant 6-386.

REFERENCES

1. Porter, P.N., R.H. Meints, and K. Mesner. 1979. Enhancement of erythroid colony growth in culture by hemin. Exp. Hematol. 7: 11.
2. Ibrahim, N.G., J.D. Lutton, and R.D. Levere. 1982. The role of haem biosynthetic and degradative enzymes in erythroid colony development: the effect of haemin. Br. J. Haematol. 50: 17.
3. Monette, F.C., and S.A. Holden. 1982. Hemin enhances the in vitro growth of primitive erythroid progenitor cells. Blood 60: 527.
4. Holden, S.A., H.N. Steinberg, E.A. Matzinger, and F.C. Monette. 1983. Further characterization of the hemin-induced enhancement of primitive erythroid progenitor cell growth in vitro. Exp. Hematol. 11: 953.
5. Lu, L., and H.E. Broxmeyer. 1983. The selective enhancing influence of hemin and products of human erythrocytes on colony formation by human multipotential (CFU_{GEMM}) and erythroid (BFU_E) progenitor

cells in vitro. Exp. Hematol. 11: 721.

6. Rothmann, J., C.F. Hertogs, Z. Malik, and D.H. Pluznik. 1983. Hemin stimulating effect on colony formation of leukemic and bone marrow cells. Exp. Hematol. 11: 147.

7. Stenzel, K.H., A.L. Rubin, and A. Novogrodsky. 1981. Mitogenic and co-mitogenic properties of hemin. J. Immunol. 127: 2469.

8. Chen, J-J., and I.M. London. 1981. Hemin enhances the differentiation of mouse 3T3 cells to adipocytes. Cell 26: 117.

9. Ishii, D.N., and G.M. Maniatis. 1978. Haemin promotes rapid neurite outgrowth in cultured mouse neuroblastoma cells. Nature 274: 372.

10. Verger, C., S. Sassa, and A. Kappas. 1983. Growth-promoting effects of iron- and cobalt-protoporphyrins on cultured embryonic cells. J. Cell Physiol. 116: 135.

11. Rutherford, T.R., J.B. Clegg, and D.J. Weatherall. 1979. K562 human leukaemic cells synthesise embryonic haemoglobin in response to haemin. Nature 280: 164.

12. Ganguly, S., A.I. Skoultchi, M.D. Garrick, L.M. Garrick, A.S. Campbell, and B.P. Alter. 1988. Differential synthesis of β-major and β-minor globin proteins in murine erythroleukemia cells is regulated at the transcriptional level. J. Biol. Chem. 263: 3216.

13. Bonanou-Tzedaki, S.A., M. Sohi, and H.R.V. Arnstein. 1981. Regulation of erythroid cell differentiation by haemin. Cell Differ. 10: 267.

14. Bonanou-Tzedaki, S.A., M.K. Sohi, and H.R.V. Arnstein. 1984. The effect of haemin on RNA synthesis and stability in differentiating rabbit erythroblasts. Eur. J. Biochem. 144: 589.

15. Bruns, G.P., and I.M. London. 1965. The effect of hemin on the synthesis of globin. Biochem. Biophy. Res. Commun. 18: 236.

16. Zucker, W.V., and H.M. Schulman. 1968. Stimulation of globin-chain initiation by hemin in the reticulocyte cell-free system. Proc. Natl. Acad. Sci. USA 59: 582.

17. Adamson, S.D., E. Herbert, and S.F. Kemp. 1969. Effects of hemin and other porphyrins on protein synthesis in a reticulocyte lysate cell-free system. J. Mol. Biol. 42: 247.

18. Muller-Eberhard, U., J. Javid, H.H. Liem, A. Hanstein, and M. Hanna. 1968. Plasma concentrations of hemopexin, haptoglobin and heme in patients with various hemolytic diseases. Blood 32: 811.

19. Kaye,F.J.,R. Weinberg, J. Schofield, and B.P. Alter. 1987. The effect of hemin in vitro and in vivo on human erythroid progenitor cells. Int. J. Cell Cloning 5: 74.

20. Weinberg, R.S., J.D. Goldberg, J.M. Schofield, A.L. Lenes, R. Styczynski, and B.P. Alter. 1983. Switch from fetal to adult hemoglobin is associated with a change in progenitor cell population. J. Clin. Invest. 71: 785.

21. Rinehart, J.J., E.D. Zanjani, B. Nomdedeu, B.J. Gormus, and M.E. Kaplan. 1978. Cell-cell interaction in erythropoiesis. Role of human monocytes. J. Clin. Invest. 62: 979.

22. Weinberg, R.S., S.E. Antonarakis, H.H. Kazazian, Jr., G.J. Dover, S. H. Orkin, A.L. Lenes, J.M. Schofield, and B.P. Alter. 1984. Fetal hemoglobin synthesis in erythroid cultures in hereditary persistence of fetal hemoglobin and β⁰-thalassemia. Blood 63: 1278.

23. Snedecor, G.W., and W.G. Cochran. 1980. Statistical Methods, 7th Ed. Iowa State University Press, Ames Iowa. 507 pages.

24. Grasset, M-F., and J.P. Blanchet. 1984. Erythroid precursors cultured from adult mice are sensitive to H_2O_2 toxicity. In Vitro 20: 302.

25. Javid, J., and P.K. Pettis. 1983. Fetal hemoglobin accumulation in vitro. Effects of adherent mononuclear cells. J. Clin. Invest. 71: 1356.

THE EFFECT OF DIMETHYL SULFOXIDE ON HEME SYNTHESIS AND THE

ACUTE PHASE REACTION IN HUMAN HepG2 HEPATOMA CELLS

Shigeru Sassa, Fuyuki Iwasa, and Richard Galbraith

The Rockefeller University
New York, NY 10021

Heme is an essential prosthetic group for a variety of hemeproteins which are involved in mitochondrial electron transport, microsomal cytochrome P-450-dependent mixed function oxidations and oxygen transport. δ-Amino-levulinic acid (ALA) synthase [succinyl CoA:glycine C-succinyl-transfer-ase(decarboxylating)] (EC 2.3.1.37) is the first enzyme of the heme bio-synthetic pathway and catalyzes the condensation of glycine and succinyl CoA to form ALA. ALA synthase activity in the normal rat liver is very low, and is rate-limiting for heme formation[1]. In animal and avian liver cells, the enzyme level is repressed by heme and increased by treatment of animals with porphyrogenic chemicals such as 2-allyl-2-isopropylacetamide and 3,5-diethoxycarbonyl-1, 4-dihydrocollidine. In contrast to the liver, the regulation of ALA synthase in other tissues appears to be different, and the enzyme activity may not always be rate-limiting for heme formation in non-hepatic tissues, e.g., in erythroid cells[4-8]. Little is known about the regulation of ALA synthase and heme synthesis in human liver. For example, it is not known whether the level of ALA synthase in human liver is under feedback control by heme. This is largely because no satisfactory isolated human liver cell preparations have been available for such studies.

HepG2 cells, a human hepatoma cell line, are morphologically similar to parenchymal cells of the liver and synthesize major plasma proteins and receptors for a variety of hormones and growth factors[9]. Thus HepG2 cells are the best available model to further our understanding of the mode of regulation of heme biosynthesis in human liver cells. In this paper, we describe that HepG2 cells maintain active heme synthesis and that heme synthesis is increased in response to the treatment of cells with dimethyl sulphoxide (DMSO). DMSO treatment of cells also increases the activity of ALA synthase, and ALA dehydratase, and also elicits an acute phase-like reaction.

MATERIALS AND METHODS

Materials

DMSO (spectro grade) was purchased from Fisher, Fair Lawn, NJ; all tissue culture materials were from GIBCO, Grand Island, NY. Polybenzimida-zole Aurorez resin was provided by Hoeschst Celanese, Charlotte, NC. Rabbit anti-(human protein) antisera and IgG purified from anti-serum were purchased either from Behring Diagnostics, La Jolla, CA, or DAKO

Molecular Biology of Erythropoiesis
Edited by J. L. Ascensao *et al.*
Plenum Press, New York

Corp., Santa Barbara, CA. Standard human proteins were from Cappel, Westchester, PA (albumin) or Behring (fibrinogen, transferrin, α-fetoprotein (AFP), and haptoglobin). Recombinant human interleukin (IL)-6 was generously provided by Dr. S. Clark, Genetics Institute, Cambridge, MA. Agarose was purchased from Bio-Rad, Richmond, CA, and sodium barbital was from Fisher. Other chemicals were of reagent grade and obtained from Sigma Chemical Co., St. Louis, MO.

Cell Culture

HepG2 cells were kindly provided by Dr. Barbara Knowles of the Wistar Institute, Philadelphia, PA. Cells were routinely grown in Corning 100x20 mm tissue culture dishes (Corning, NY), in minimum essential medium with Earles salts supplemented with 10%(v/v) fetal bovine serum, 100 units of penicillin/ml, 100 μg of streptomycin/ml and 2 mM glutamine. Routine subcultures were made weekly at 1:3 dilutions, which maintained cells in the exponential growth phase; cells were fed with fresh medium every 2-3 days. Additions of chemicals, e.g., DMSO, were made directly to the flasks. Cultures were incubated in the dark and cells were harvested by treatment with 0.05%(w/v) trypsin containing 0.53 mM EDTA for 5 min at 37°C, washed and resuspended in Earles salts and, after 15 passages through a pipette to break up clumps, their numbers were counted in a model F_N Coulter counter.

Preparation of Chemical Solutions

Hemin was prepared at a concentration of 20 mM in 0.2M KOH/methanol (1:1, v/v), and diluted 10 times with culture medium, as described previously[4]. The solutions were then filtered through a Millipore membrane (0.22 μm pore) for sterilization. Other chemicals were dissolved in sterile water at a 100-fold greater concentration than the final concentration in the culture medium.

Plasma Protein Assays

Albumin and AFP synthesized and secreted by HepG2 cells into culture media were quantified by rocket immunoelectrophoresis[10]. For determining fibrinogen and haptoglobin, culture media were dialyzed against 10mM-sodium phosphate buffer, pH 7.4, and concentrated (10-fold) prior to rocket immunoelectrophoresis, by using a Speed-Vac concentrator (Savant, Hicksville, NY). Plasma protein synthesis was expressed as μg of protein secreted per day using mg of cellular protein.

ALA Synthase Assay

ALA synthase activity of HepG2 cells was determined using a radiochemical assay described by Brooker et al.[11] with the following modifications; succinylacetone was included in the reaction mixture to ensure complete inhibition of ALA dehydratase activity[12], and polybenzimidazole resin, instead of ethyl acetate extraction, was used for rapid and easy recovery of 2-methyl-3-acetyl-4-(3-propionic acid) pyrrole. Cells from a 100 mm dish were collected in a 15 ml polypropylene conical centrifuge tube (Sarstedt, Princeton, NJ). The cell pellet was resuspended in 90 μl of a mixture containing 10 μM pyridoxal 5'-phosphate, 0.1 mM (CaMg)EDTA, 0.1%(w/v) Triton X-100, and 5 mM Tris-HCl, pH 7.4. An aliquot of 90 μl was added to a glass tube (12x75 mm) for the ALA synthase assay and another aliquot of 20 μl was used for the protein assay.

Other Assays

ALA dehydratase activity was determined using 10^6 cells per assay according to the method described previously[8]. Heme content was determined fluorometrically using 10^5 cells per assay[8]. Protein concentrations were determined according to the method of Lowry et al.[13], after digestion of cells in 1M sodium hydroxide and dilution to 2 ml so that the concentration of Triton (<0.0001 %[w/v]) did not interfere with the assay. Data were expressed based on mg protein per assay. Statistical analyses of data were performed by using Student's unpaired one-tail t test.

RESULTS

Acute Phase Reaction

When HepG2 cells were incubated with DMSO, both albumin and AFP synthesis were inhibited, while the synthesis of haptoglobin was significantly increased (Figure 1A). Similar changes were also elicited by treatment of cells and LPS-conditioned macrophage medium (Fig. 1B), or with rhIL-6 (Fig. 1C), a major acute phase inducer.

Heme Synthesis

Treatment of HepG2 cells with 2%(v/v) DMSO for 96 h resulted in a significant reduction in cell number (ca. 40%) without alterations in the protein content per cell. DMSO treatment also showed small, but significantly ($p < 0.01$) higher heme content than untreated cells (Table I).

ALA Dehydratase

HepG2 cells contained measurable ALA dehydratase activity. In addition, when HepG2 cells were treated with 2%(v/v) DMSO, cells showed a significant increase in ALA dehydratase activity. Increases in enzyme activity were dose- and time-dependent, and the maximal increase in enzyme activity was seen at 2%(v/v) DMSO and at 48 h or thereafter (Fig. 2A and 2B). These changes in ALA dehydratase activity were very similar to those elicited in MEL cells by DMSO[8], but strikingly different from those in liver cells maintained in culture for extended time periods[14].

In order to differentiate whether the increased ALA dehydratase activity in DMSO-Treated HepG2 cells was due to an increased catalytic activity of the enzyme, or due to increased synthesis of the enzyme, we examined both enzyme activity and immunoquantifiable enzyme concentration in HepG2 cells. Treatment of HepG2 cells with DMSO increased both the activity and the amounts of enzyme protein of ALA dehydratase and the increases were comparable. α-Amanitin, a specific inhibitor of RNA polymerase II, inhibited the DMSO-mediated increases in ALA dehydratase activity and the enzyme protein, while it did not affect the enzyme activity and the enzyme protein concentration in untreated cells (Fig. 3). Thus the proportional changes in enzyme activity and enzyme protein concentration demonstrate that the changes in ALA dehydratase activity in HepG2 cells following treatment with these chemicals are due to de novo synthesis of the enzyme. Since the DMSO-mediated increase in ALA dehydratase was blocked by α-amanitin, while the inhibitor did not affect the level of ALA dehydratase in untreated cells, the increased de novo synthesis of ALA dehydratase after DMSO treatment must result from transcriptional activation of the gene.

The transcriptional induction of ALA dehydratase was also demonstrated in experiments in which cells were treated with 5-bromo-2'-deoxyuridine (BrdU). BrdU is a thymidine analogue and is known to be efficiently incorporated into DNA and to suppress the expression of differentiated characteristics of many cell types[8]. When HepG2 cells were treated with

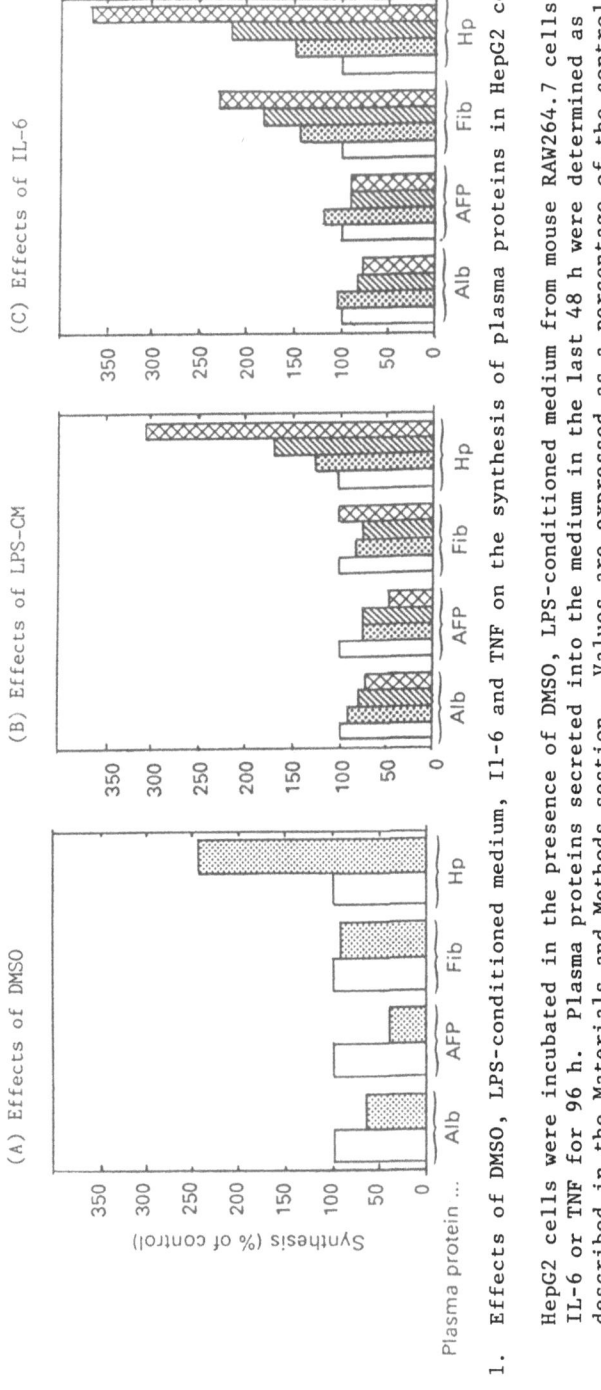

Fig. 1. Effects of DMSO, LPS-conditioned medium, Il-6 and TNF on the synthesis of plasma proteins in HepG2 cells.

HepG2 cells were incubated in the presence of DMSO, LPS-conditioned medium from mouse RAW264.7 cells, IL-6 or TNF for 96 h. Plasma proteins secreted into the medium in the last 48 h were determined as described in the Materials and Methods section. Values are expressed as a percentage of the control value. (A) Effects of DMSO; (B) effects of LPS-conditioned medium; (C) effects of IL-6. Control values were as follows: albumin (Alb), 20-37 μg/day per mg cell protein; AFP, 3.1-17 μg/day per mg protein; fibrinogen (Fib), 1.0-8.5 μg/day per mg protein, and haptoglobin (Hp), 0.03-0.20 μg/day per mg protein.

(A) Effects of DMSO: ☐ , Control: ▨ , 2%(v/v) DMSO.

(B) Effects of LPS-conditioned medium (LPS-CM): ☐ , Control; ▨ , 0.05%(v/v) ,
0.25%(v/v); ▨ , 0.5%(v/v) LPS-CM.

(C) Effects of IL-6: ☐ , Control; ▨ , 1U/ml; ▨ , 20 U/ml; ▨ , 200 U/ml.

Values are the means of duplicate determinations from single experiments (A and B) or two experiments (C).

Table I. Effects of DMSO on the growth and the heme content of HepG2 Cells.

Treatment	Incubation period (h)	Cell number per 6cm dish (x 10^{-6})	Protein content (mg/10^6cells)	Heme content (nmol/10^6cells)
		Mean ± S.E.	Mean ± S.E.	Mean ± S.E.
None	96	7.00 ± 0.18	0.34 ± 0.03	28.3 ± 3.03
DMSO, 2%(v/v)	96	4.81 ± 0.59 (p<0.001)	0.35 ± 0.05	36.1 ± 3.37 (p<0.01)

Each datum point represents the mean ± S.E. of 6 dishes.

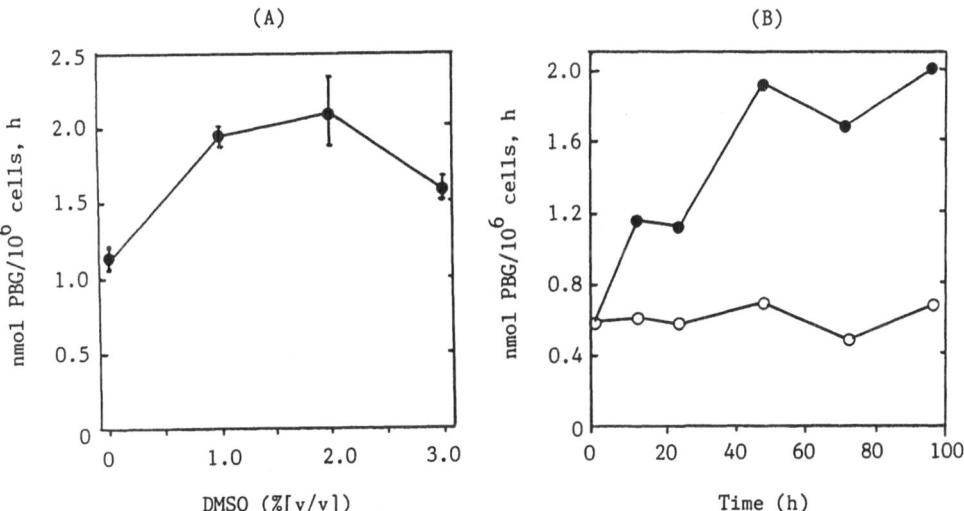

(A)   (B)

DMSO (%[v/v])   Time (h)

Fig. 2.   The effect of DMSO on ALA dehydratase activity in HepG2 Cells.

(A) HepG2 cells were incubated with various concentrations
of DMSO as indicated on the abscissa for 96 h.   (B) Cells
were incubated with (·) or without (·) 2%(v/v) DMSO from
zero time and harvested at the indicated times. All cells
were fed with fresh media with or without DMSO after 72 h.
Cell numbers were counted and ALA dehydratase activities
were determined as described in the Materials and Methods
section. Data are means ° S.E. of duplicate or triplicate
assays. All data points of the DMSO-treated group were
significantly greater than those of the untreated control
(A), or the zero time control (B), (p<0.001).

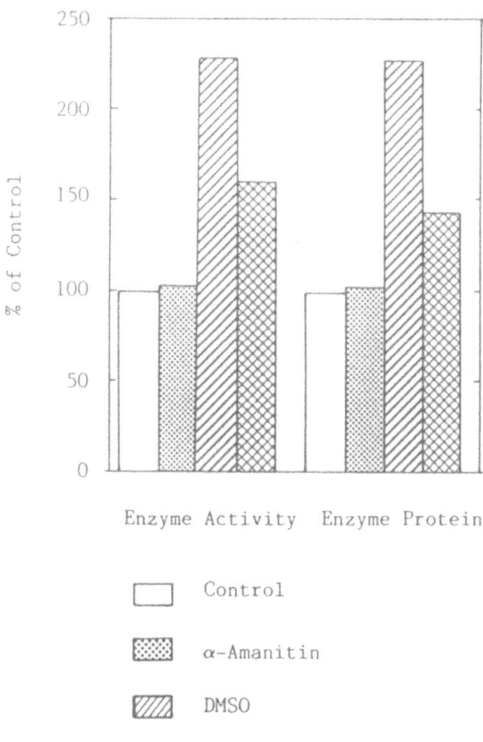

Fig. 3.   The effect of DMSO and α-amanitin
          on the activity and the enzyme
          protein of ALA dehydratase in
          HepG2 cells.

          HepG2 cells were incubated for 96 h,
          with or without 2%(v/v) DMSO, and
          0.1μg of α-amanitin/ml. Determinations
          of ALA dehydratase activities and pro-
          tein concentrations were made in triplicate
          and expressed as percentages of the un-
          treated control. Enzyme activity and
          protein concentrations in untreated cells
          were 0.74 nmol PGB/10^6 cells respectively.

20 μM BrdU, there was no effect on ALA dehydratase activity in control
cells, but the DMSO-mediated increase in ALA dehydratase was suppressed.
When 70 μM thymidine was added simultaneously with BrdU and DMSO, the
BrdU-mediated suppression of the ALA dehydratase induction was abolished
(Fig. 4).

ALA Synthase

     The activity of ALA synthase was also detected in untreated HepG2
cells. ALA synthase activity in untreated HepG2 cells was considerably
lower than the level reported for rat liver[11], biopsy samples of human
liver[16-18], or cultured human lymphocytes[19], but comparable to those
reported for cultured fibroblasts[20,21]. Treatment of HepG2 cells with
2%(v/v) DMSO for up to 48 h did not influence ALA synthase activity,
but it increased the enzyme activity in cells treated for 72 h or longer

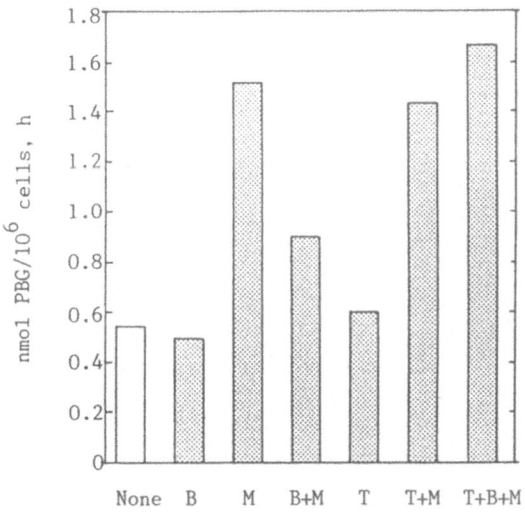

Fig. 4. The effect of bromodeoxyuridine on the
         induction of ALA dehydratase by DMSO in
         HepG2 cells.

         HepG2 cells were treated with the in-
         dicated chemicals at the following
         final concentrations: 5-brom-2'-
         deoxyuridine (B) 20 µM DMSO (M) 2%
         (v/v), and thymidine (T) 70 µM. Media
         with respective additions were replaced
         after 72 h. Cells were harvested at 96 h
         under subdued light and ALA dehydratase
         activities were determined as described
         in the Materials and Methods section.
         Data are the mean ± S.E. of triplicate
         dishes.

(Fig. 5). IL-6 by itself had little effect on ALA synthase, but it poten-
tiated the increase in ALA synthase activity induced by DMSO (Fig. 5).

    ALA synthase activity in HepG2 cells was markedly suppressed when
cells were treated with hemin in culture. A half-maximal inhibition
of the enzyme activity in cultured cells was observed at a hemin concentration
of 1 µM (Table II). In contrast, the enzyme activity was unaffected
by hemin, even at a concentration of 100 µM, if hemin was added <u>in vitro</u>
to the enzyme preparation (data not shown). These findings indicate
that the observed enzyme inhibition by hemin in cultured cells is not
due to enzyme inhibition, but rather due to inhibition of enzyme synthesis.

    The half-life of ALA synthase activity in cells after hemin treatment
was approximately 2 h (Table III). This half-life was similar to that
obtained after treatment of cells with cycloheximide, an inhibitor of
protein synthesis, but was considerably shorter than that observed with
actinomycin D (Table III).

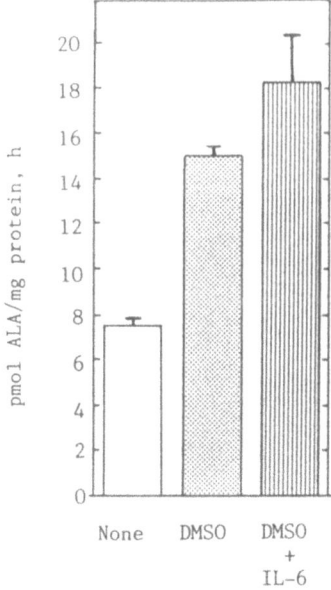

Fig. 5. The effect of DMSO and IL-6
on ALA synthase activity
in HepG2 cells.

HepG2 cells were incubated
in the presence of 2%(v/v)
DMSO and/or IL-6 (200 U/ml)
for 72 h. ALA synthase
activities were determined
by a radiochemical method
as described in the Materials
and Methods section. Data
are means ± S.D. of triplicate
dishes. $p < 0.001$ between DMSO
and control. $p < 0.05$ between
DMSO and DMSO+IL-6.

DISCUSSION

Liver cells are unique in that they synthesize a number of plasma
proteins and exhibit the acute phase reaction in response to inflammation
and toxic stimuli[22]. The acute phase reaction is characterized by increased
synthesis of certain plasma proteins and decreased synthesis of albumin.
Another unique feature of the liver is the ability to synthesize a large
amount of heme for incorporation into the microsomal cytochrome P450s
which are involved in the metabolism of a large variety of compounds
such as drugs, chemicals, endogenous steroids, fatty acids, prostaglandins
and vitamin D[23]. Thus it is possible that a major change(s) in heme
synthesis may occur if the function of the liver is significantly altered
during the acute phase reaction. Little is known, however, about the
influence of the acute phase reaction on heme biosynthesis in the liver.
The results of the present study demonstrate that incubation of HepG2
cells with LPS-conditioned medium of macrophages, or IL-6 stimulates
the synthesis of haptoglobin (Fig. 1B & 1C). Treatment of cells with
LPS-conditioned medium was accompanied by decreased synthesis of albumin

110

Table II. Effects of hemin on ALA synthase activity in HepG2 cells.

Treatment		ALA synthase activity		
		pmol/mg protein, h	% of control	P value
		Mean ± S.D. (n=2)	Mean	
None		11.94 ± 0.95[#]	100	
Hemin,	0.5μM	9.54 ± 0.30	86	n.s.
Hemin,	1.0μM	6.28 ± 1.68	56	<0.005
Hemin,	5.0μM	3.17 ± 0.03	28	<0.001
Hemin,	10.0μM	2.58 ± 0.58	23	<0.001
Hemin,	50.0μM	1.17 ± 0.20	11	<0.001
Hemin,	100.0μM	1.01 ± 0.16	9	<0.001

: n = 5
P value is for the difference between the sample and the control.
n.s. : not significant

Cells were incubated with various concentrations of hemin for 6 h,
prior to the determination of ALA synthase activity.

and AFP. Both these positive and negative changes in the synthesis of certain plasma proteins are characteristic of the acute phase reaction[22]. In addition, treatment of cells with DMSO induced an acute phase-like reaction as characterized by increased synthesis of haptoglobin, and decreased synthesis of albumin and AFP. These findings indicate that the acute phase reaction can be induced in HepG2 cells by treatment with LPS-conditioned medium, IL-6 or DMSO.

Treatment of cells with DMSO resulted not only in the induction of the acute phase reaction, but also in increased activities of ALA synthase and ALA dehydratase, and an increased heme content. Increases in ALA dehydratase activities were shown to be due to de novo induction of enzyme synthesis mediated at the level of transcription. It should be noted that ALA dehydratase activity is known to fall precipitously and disappear rapidly in rat hepatocytes after establishment of cells in primary culture[14]. ALA synthase activity is also refractory to induction in primary cultures of chick embryo hepatocytes after 3 days[24]. Heme content increased after 72 h (data not shown), when ALA synthase activity also started to increase.

The level of ALA synthase in HepG2 cells, but not the enzyme activity itself, is subject to inhibition by hemin treatment, suggesting that enzyme synthesis is regulated by feed-back repression by hemin. The half-life of ALA synthase after hemin or cycloheximide treatment was approximately 2 h, which was considerably shorter than the half-life observed with actinomycin D. These findings suggest that the half-life of ALA synthase mRNA may be considerably longer than that of the enzyme itself, and that the effect of hemin on ALA synthase synthesis may be

Table III. Half-lives of ALA synthase activity in HepG2 cells
treated with hemin, cycloheximide and actinomycin D.

Treatment	Half-life of ALA synthase
Hemin	2 h
Cycloheximide	2 h
Actinomycin D	12 h

Hemin (10 μM), cycloheximide (1 μg/ml) or actinomycin D (0.25 μg/ml) was added to cultures at time 0. Cells were harvested at 0, 1, 2, 3.5, 5, 7, 9 and 12 h for ALA synthase assays.

exerted at a post-transcriptional level. Half-lives of ALA synthase in rat liver[25] and chick embryo hepatocytes[26] have been reported to be similar to or longer than that of the enzyme itself.

We have reported earlier that HepG2 cells retain measurable activities of porphobilinogen deaminase[27], uroporphyrinogen decarboxylase[27], cytochrome P450[28] and mixed function oxidase activities[28]. Additionally, we have shown that there is a dose-dependent increase in 7-ethoxycoumarine o-de--ethylase activity in HepG2 cells after treatment with DMSO, while aryl hydrocarbon hydroxylase activity was inhibited by the same treatment[28]. These findings indicate that HepG2 cells maintain active heme synthesis, and that heme synthesis is stimulated when cells undergo the acute phase reaction. Studies on the effect of DMSO and acute phase inducers on enzymes in the heme catabolic pathway and cytochrome P450 isozymes are currently under way in our laboratory. Our findings also indicate that the synthesis of ALA synthase in HepG2 cells is repressed by heme in a feed-back fashion. Thus it appears that HepG2 cells may be a useful model for the study of heme synthesis in human liver-derived cells.

SUMMARY

Effects of DMSO on heme synthesis and enzymes of the heme biosynthetic pathway were examined in human HepG2 hepatoma cells. HepG2 cells contain measurable levels of ALA synthase and ALA dehydratase, and their levels are increased after treatment of cells with DMSO. DMSO treatment also led to increases in heme content and the synthesis of haptoglobin, while it decreased the synthesis of albumin and AFP. Changes in plasma protein synthesis after DMSO treatment are characteristic of those known to occur in the acute phase reaction. These findings suggest that profound changes in heme synthesis may occur during the acute phase reaction.

ACKNOWLEDGEMENT

This work was supported in part by U.S.P.H.S. grant DK-32890 and the Suntory Fund for Biomedical Research. We are grateful to Dr. Barbara Knowles for supplying us with HepG2 cells, to Dr. Jorge E. Ramirez, Hoechst Celanese, for supplying us with Aurorez resin, to Dr. H. Fujita for the determination of ALA dehydratase protein by radioimmunoassay, and to Dr. A. Kappas for his encouragement in this work. The excellent technical assistance of Ms. Luba Garbaczewski and Ms. Anna Gajewski is gratefully acknowledged.

REFERENCES

1. Kappas, A., S. Sassa, and K.E. Anderson. 1982. In: The Metabolic Basis of Inherited Disease (Stanbury, J.B., Wyngaarden, J.B., Frederickson, D.S., Goldstein, J.L. and Brown, M.S., eds.), McGraw-Hill Book Co., New York. pp. 1301-1384.
2. Marver, H.S., A. Collins, D.P. Tschudy, and J. Rechcigl, Jr. 1966. J. Biol. Chem. 241: 4323-4329.
3. Granick, S. and G. Urata. 1963. J. Biol. Chem. 238: 821-827.
4. Granick, J.L. and S. Sassa. 1978. J. Biol. Chem. 253: 5402-5406.
5. Hoffman, R., N. Ibrahim, M.J. Murnane, A. Diamond, B.G. Forget, and R.D. Levere. 1980 Blood 56-567-570.
6. Nakao, K., S. Sassa, O. Wada, and F. Takaku. 1968. Ann. N.Y. Acad. Sci. 149: 224-228.
7. Wada, O., S. Sassa, F. Takaku, Y. Yano, G. Urata, and K. Nakao. 1967. Biochim. Biophys. Acta 148: 585-587.
8. Sassa, S. 1976. J. Exp. Med. 143: 305-315.
9. Knowles, B.B., C.C. Howe, and D.P. Aden. 1980. Science 209: 497-499.
10. Grieninger, G., J. Pindyck, K.M. Hertsberg, and M.W. Mosesson. 1979. Ann. Clin. Lab. Sci. 9: 511-517.
11. Brooker, J.D., G. Srivastava, B.K. May and W.H. Elliott. 1982. Enzyme 28: 109-119.
12. Sassa, S. and A. Kappas. 1983. J. Clin. Invest. 71: 625-634.
13. Lowry, O.H., N.J. Rosebrough, A.L. Farr, and R.J. Randall. J. Biol. Chem. 193: 265-275.
14. Guzelian, P.S., L. O'Connor, S. Fernandez, U. Chan, P. Giampietro, and R. Desnick. 1984. Life Sci. 31: 1111-1116.
15. Marks, P.A. and R.A. Rifkind. 1978. Ann. Rev. Biochem. 47: 419-448.
16. Dowdle, E.B., P. Mustard, and L. Eales. 1967. South Afr. Med. J. 2: 1093-1096.
17. Strand, L.J., B.F. Felsher, A.G. Redecker, and H.S. Marver. 1970. Proc. Nat. Acad. Sci. USA 67: 1315-1320.
18. Nakao, K., O. Wada, T. Kitamura, K. Uono and G. Urata. 1966. Nature 210: 838-839.
19. Sassa, S., G.L. Zalar, and A. Kappas. 1978. J. Clin. Invest. 61: 499-508.
20. Meyer, U.A. 1973. Enzyme 16: 334-342.
21. Bonkowsky, H.L., D.P. Tschudy, E.C. Weinbach, P.S. Ebert, and J.M. Doherty. 1975. J. Lab. Clin. Med. 85: 93-102.
22. Jamieson, J.C., G. Lammers, R. Janzen, and B.M.R.N.J. Woloski. 1987. Comp. Biochem. Physiol. 878: 11-15.
23. Nebert, D.W., and F.J. Gonzales. 1987. Ann. Rev. Biochem. 56: 945-993.
24. Sardana, M.K., S. Sassa, and A. Kappas. 1982. In: Isolation Characterization and Use of Hepatocytes. (R.A. Harris and N.W. Cornell, eds.) Elsevier Biomedical, New York. pp. 111-116.
25. Tschudy, D.P., H.S. Marver, and A. Collins. 1965. Biochem. Biophys. Res. Comm. 21: 480-487.
26. Sassa, S. and S. Granick. 1970. Proc. Nat. Acad. Sci. USA 67: 517-522.

27. Galbraith, R.A., S. Sassa, and H. Fujita. 1988. <u>Biochem. Biophys.</u>
    <u>Res. Comm.</u> 153: 869-874.
28. Sassa, S., O. Sugita, R.A. Galbraith, and A. Kappas. 1987. <u>Biochem.</u>
    <u>Biophys. Res. Comm.</u> 143: 52-57.

CYTOCHROME P450 DEPENDENT ARACHIDONIC ACID

METABOLISM IN HEMOPOIETIC CELLS

J.D. Lutton, M. Laniado Schwartzman, and N.G. Abraham

Departments of Medicine and Pharmacology
New York Medical College
Valhalla, NY 10595

ABSTRACT

This report demonstrates for the first time that human peripheral blood neutrophils and human bone marrow cells metabolize arachidonic acid (AA) by a cytochrome P450 dependent mechanism. The formation of the cytochrome P450 (P450) arachidonate metabolites is dependent on the addition of NADPH, prevented by SKF-525A (100µM), which is an inhibitor of cytochrome P450 enzymes, but not affected by the addition of BW-755C, a dual inhibitor of cyclooxygenase and lipoxygenase activities. Addition of the Ca^{++}-ionophore A23187 to cell preparataions stimulated the release of both cyclooxygenase and lipoxygenase products but did not affect the formation of the P450 metabolites. Incubation of cell preparations with ^{14}C-AA yielded a P450 dependent peak which eluted at 19 min. on reverse phase HPLC and was distinct from 5 and 15-hydroxyeicosatetraenoic acids (HETE's), prostaglandins and other leukotrienes. Recovery of the P450 dependent metabolite(s) was accomplished and preparations were tested in bone marrow clonal culture in order to determine if the substance(s) has/have any effect on the growth and differentiation of bone marrow cells. Results demonstrated that some of the material possessed powerful erythroid colony (CFU-E) enhancing activity at concentrations between 10^{-8}-10^{-14}M. These results demonstrate that human bone marrow and peripheral blood neutrophils possess a third pathway for AA metabolism which is P450 dependent, and that metabolite(s) of this pathway may have potent stimulatory effects on erythropoiesis.

INTRODUCTION

Bone marrow is the source of hemopoietic progenitor cells which arise from pluripotent stem cells within the hemopoietic mileau[1]. As with the liver, the bone marrow carries on a multitute of life dependent functions and disturbances in hemopoietic equilibria may be the consequence of/or cause of multiple clinical disorders. The generation of regulatory growth substances, interleukins, cytokines and various mediators of the inflammatory response are only a few of the many important factors that are derived from bone marrow and hemopoietic cells[1]. Recently it has been demonstrated that the bone marrow may function as an important drug metabolizing organ[2-3] and we have characterized a cytochrome P450 (P450) metabolizing system in _in vitro_ hemopoietic bone marrow erythroid colonies (CFU-E) and myeloid colonies (CFU-GM)[4].

Prostaglandins and leukotrienes are known to have major effects on a variety of hematologic processes and the generation of these substances is a complex process involving the metabolism of arachidonic acid (AA) via cyclooxygenase and lipoxygenase[5]. Recently, a third pathway for the metabolism of AA which is a NADPH dependent P450 system have been characterized[6-7]. This system has since been demonstrated in a variety of tissues including human liver, kidney, cornea, and some metabolic products of this pathway have been isolated and characterized[7-11]. These products have marked effects on blood vessel tone, platelet aggregation, ion transport mechanisms, membrane premeability, and the activity of Na^+-K^+ ATPase[8-11].

Little is known about the role of the bone marrow and progenitor cells in the metabolism of AA and the significance of P450 in these cells. Therefore, the present study was undertaken and demonstrates that human bone marrow and peripheral blood neutrophils possess a P450 dependent pathway for the metabolism of AA. Additionally, products of this pathway have potent effects on hemopoietic growth and may be of significance for modulating normal/abnormal hematopoiesis.

METHODS

Cell Preparations

Peripheral blood: Heprinized human peripheral blood (20-50ml) was obtained from normal volunteers. Peripheral blood neutrophils were prepared by separation on Ficoll-Hypaque density centrifugation (Ficoll-Paque, Pharmacia), and then removal of monocytes-macrophages by adherence[1].

Bone marrow: Heprinized human bone marrow was obtained by aspiration from the posterior iliac crest of normal volunteers. In all cases, informed consent was obtained. Buffey coat cells were separated on Ficoll-Hypaque and then non-adherent cells obtained by removal of adherent cells. All cell preparations were washed and resuspended in Iscoves modified Dulbeccos medium (IMDM). In most cases, 10^6 cells were used for HPLC assays, and 6×10^5/ml for colony cultures. When murine cells were to be used for assays, bone marrow cells were flushed from the femus of DBA mice, washed and then resuspended to the desired concentration.

Hemopoietic Colony Assays

Murine and human bone marrow erythroid colonies (CFU-E) were grown in methylcellulose similar to procedures described previously[1]. CFU-E were cultured in the presence of 0.4-1.0 μ/ml erythropoietin (Epo, Toyobo), and then scored after 3-7 days of culture at 37°C. Metabolites to be assayed were included in CFU-E culture medium in the presence and absence of Epo.

Arachidonic Acid and Cytochrome P450 Metabolisms

Cells were incubated with 7μM ^{14}C-AA acid for 30 min. at 37°C in the presence or absence of A23187 (5μM), SKF-525A (100μM) or BW 755C (100μM). The reaction was terminated by acidification to pH 4.0 and AA metabolites were extracted with ethyl acetate. Determination of radioactive AA acid metabolites was performed by separation on reverse phase HPLC and their migrations compared to standards. These procedures have been described in detail previously[7-8]. HPLC was performed on a C18 Microsorb column (Rainin Instruments, Woburn, MA) using a linear gradient of 1.25%/min from acetonitrile: H_2O: acetic acid (50:50:0.1) toacetonitrile: acetic acid (100:1) and the radioactivity monitored by a flow detector (Radiometric Instrument and Chemical, Tampa, FL.). For recovery of the metabolites, fractions containing the P450 AA metabolites (19 min) were pooled, evaporated, resuspended in methanol and further purified by HPLC. The purified AA metabolites were stored in ethanol under nitrogen at $-70°C$.

The metabolic inhibitors BW-755C (100µM), SKF-525A (100µM) and calcium ionophore A23187 (5µM) were from Burroughs Welcome, Inc. (Research Triangle, N.C.), Smith, Kline and French (Patterson, N.J.) and Sigma Chem. Co., (St. Louis, MO.) respectively. $[1-^{14}C]$ arachidonic acid (56 µCi/mmol) was obtained from Amersham Corp.

RESULTS

Preliminary studies indicated that the optimal concentration of ^{14}C-AA for product formation was 7-10µM. HPLC results from incubation of peripheral blood neutrophils incubated with ^{14}C-AA are represented on Figures 1 A-C. As shown in Fig. 1A, incubation of cells (neutrophils) with ^{14}C-AA (7µM) resulted in the generation of several oxygenated AA metabolites with retention time intervals varying between 6-25 minutes. Note on this graph that there is a product generated which appears at approximately 19 min. and this is distinct from the metabolites previously identified as 5 and 15-hydroxyeicosatetraenoic acids (HETES) (22-24 min) and prostaglandins (6-12 min). The product which appears at 19 min. corresponds to AA metabolites generated by P-450. This will become more apparent from the next series of experiments since inhibition or stimulation of cyclooxygenase and lipoxygenase pathways is independent of P450.

Figure 1B represents HPLC results from neutrophiles that were exposed to 100 µM SKF-525A, which is an inhibitor of P450 enzymes. Note that no 19 min. product is formed and the formation of HETE's and prostaglandins are actually enhanced. In contrast, exposure of cells to BW-755C, which inhibits lipoxygenase and cyclooxygenase, was found to have no inhibitory effect on the formation of the 19 min. product(s) (Fig. 1-C). However, formation of lipoxygenase pathway products (HETE's) was completely abolished and prostaglandin formation reduced (Fig. 1-C). Therefore these results suggest that the product(s) at the 19 min. peak is/are P450 dependent and not the result of lipoxygenase or cyclooxygenase pathways. Generation of the 19 min. retention product(s) (P450 dependent) was also obtained with human bone marrow cells (Fig. 1D). However, the extent of P450 dependent product formation appeared to vary between bone marrow samples. Figure 2-A demonstrates results from another control bone marrow sample and it can be seen that there is no obvious peak which corresponds to P450 dependent metabolites, whereas prostaglandin metabolites are obvious. Appearance of the P450 dependent peak occurred when cyclooxygenase and lipoxygenase pathways were inhibited with BW-755C (Fig. 2B). Conversely, stimulation of lipoxygenase and cyclooxygenase pathways with the calcium ionophase A23187 resulted in a complete dimmunition of P450 product(s) and enhancement of HETE(s) and prostaglandin formation (Fig. 2C).

In the next series of experiments, material was extracted from the P450 dependent fraction and then tested for its effects on in vitro erythro-poiesis. Fractions were collected, pooled, repurified and then tested at various concentrations in bone marrow erythroid CFU-E cultures. Results are presented on Table 1 and demonstrate that concentrations of the material ranging between 10^{-8}-$10^{-14}M$ had marked enhancing effects on CFU-E growth. The stimulatory effect was apparent for both human and murine bone marrow cultures and results on Table 1 are given as % control growth. Human

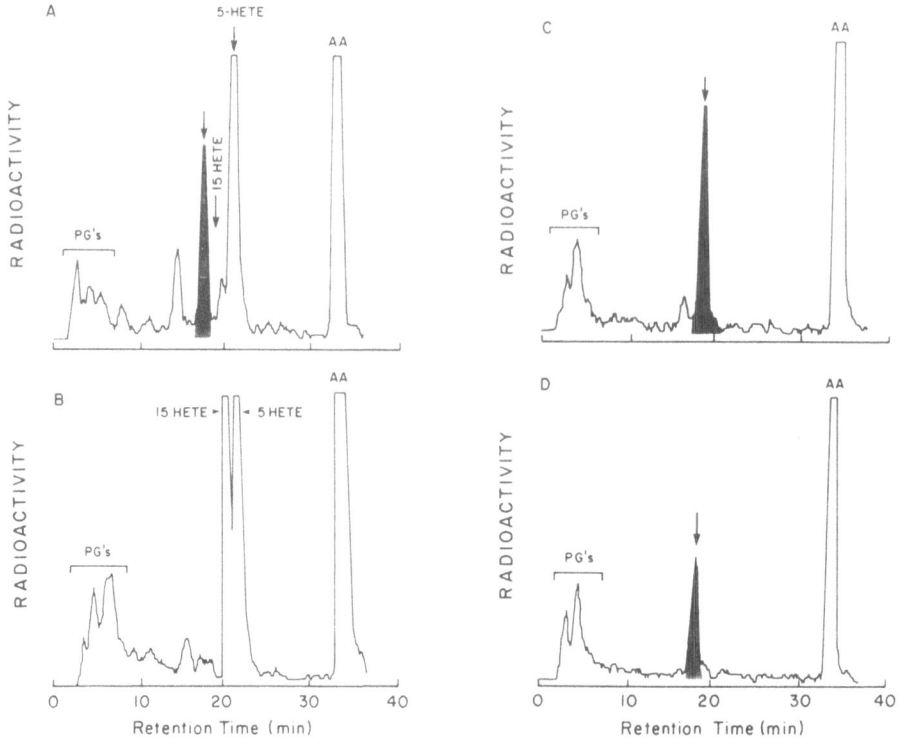

Fig. 1. Reverse-phase HPLC separation of AA metabolites formed by human
neutrophils and non-adherent bone marrow cells.

    A)   Incubation of 1 x 10^7 neutrophils with ^{14}C-AA (7 μM)
          for 30 min at 37°C;

    B)   Same as A but with 10 min preincubation with SKF-525A
          (100 μM);

    C)   Same as A but with 10 min preincubation with BW-755C
          (100 μM);

    D)   Incubation of 4 x 10^6 bone marrow cells with ^{14}C-AA
          for 30 min at 37°C.

The reaction was terminated by acidification, and radio-labeled
metabolites were extracted and separated by HPLC.

marrow cultures were very responsive and 10^{-14}M of the compound stimulated
culture colony growth equal to 316% of control growth. Murine bone marrow
appeared to be less sensitive with 10^{-8}M generating CFU-E growth equal
to 214% of control growth.

DISCUSSION

     Previously, we have demonstrated that bone marrow cells and bone
marrow hemopoietic colonies possess an active P450 metabolizing system[4].
Results from the present study clearly demonstrate that bone marrow cells
and peripheral blood neutrophils metabolize AA via P450 dependent pathway
in addition to cyclooxygenase and lipoxygenase. By employing HPLC techniques,
products of this pathway were clearly distinguishable and eluted separately
from products of lipoxygenase and cyclooxygenase including HETE's and
prostaglandins. Results with inhibitors in the present studies lend
further support for the P450 dependency of the AA metabolite(s) found
in the 19 min. band. Product formation by the P450 pathway was inhibited

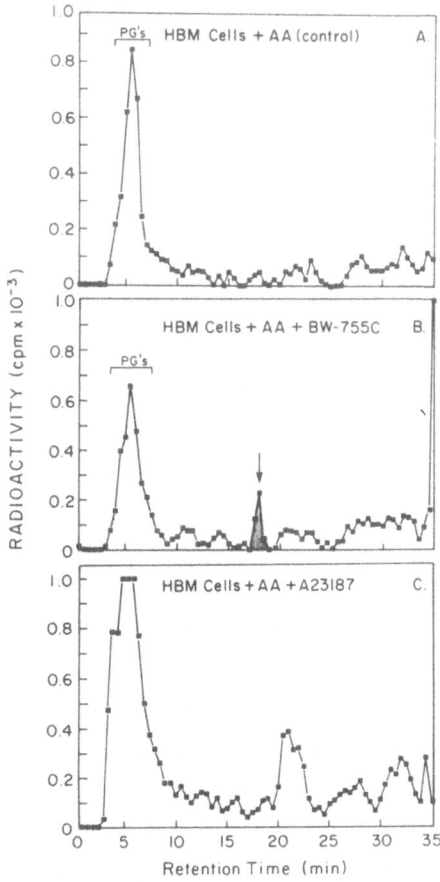

Fig. 2. Reverse-phase HPLC separation of AA
metabolites formed by human bone
marrow cells. ^{14}C-Arachidonic acid
(7 μM) was incubated for 30 min at
37°C in the presence or absence of
BW-755C (100 μM) or Ca-Ionosphore
A23187 (10 μM).

by the P450 inhibitor, SKF-525A[12], but not by BW-755C, a dual inhibitor
of cyclooxygenase and lipoxygenase[13]. Additionally, cyclooxygenase and
lipoxygenase product formation was enhanced in the presence of A23187
without any demonstratable effect on metabolite formation by the P450
system. In fact, enhancement of cyclooxygenase and lipoxygenase pathways
may actually shunt major metabolism of AA via these pathways thereby
reducing the presence and demonstratibility of the third pathway.

Results suggest that the activity of the P450 pathway might vary
between samples of bone marrow cells. It is possible that the activity
of cyclooxygenase and lipoxygenase pathways may predominate in some samples
and therefore promote a masking or depression of the third pathway. Therefore it is essential that experiments are done with metabolic inhibitors
and enhancers in order to clarify this point. Bone marrow is a complex
tissue with variable cell populations which may further complicate the
direction of metabolizing systems. However, neutrophils, which are present
in bone marrow, appear to have significant activity for the third pathway.
The precise source of metabolizing cells within the bone marrow is not

119

Table 1. Effect of a Cytochrome P450-Arachidonic Acid
Metabolite on Human and Murine Bone Marrow
CFU-E Growth*

Concentration (M)	% Control CFU-E Growth				
	10^{-14}	10^{-12}	10^{-11}	10^{-10}	10^{-8}
Marrow Sample					
Human	316	279	196	230	127
Mouse	159	161	194	197	214

* CFU-E were grown in methylcellulose cultures in the presence of
erythropoietin (Epo). Controls for human marrow cultures had 1 U Epo/ml,
and murine cultures had 0.4 U Epo/ml.

clear and certainly multiple cell populations may be involved. In this
context, both erythroid and myeloid bone marrow colony progenitor cells
were found to exhibit P450 dependent arylhydrocarbon hydroxylase activity[4].

Schwartzman and associates have uniquely demonstrated P450 dependent
AA metabolism by microsomes and cellular extracts[7-8]. Their studies
include characterization of at least two biologically active oxygenated
products that are generated by the P450 pathway[7-11]. One of these products
has potent inhibitory effects on Na^{+}-K^{+}ATPase and thus could be of great
significance to a multitude of biological processes[8]. In the present
study, a preparation of the P450 AA metabolite(s) was tested for possible
effects on in vitro bone marrow erythropoiesis. Varying concentrations
of the material were tested (10^{-7}-10^{-14}M) in cultures of human and murine
bone marrow cells. It was found that the substance(s) had potent stimulatory effects on erythroid CFU-E growth in the presence of erythropoietin
(Epo). In fact, more than three fold increases in CFU-E numbers were
obtained with as little as 10^{-14}M of the material. Thus, this naturally
occurring metabolite)s) may have powerful regulatory effects on erythropoiesis in situ under certain conditions. The significance of the P450
pathway for AA metabolism in normal and abnormal hemopoietic cells is
currently being evaluated in our laboratory.

SUMMARY

1. In addition to cyclooxygenase and lipoxygenase, arachidonic acid
    (AA) can be metabolized via a third pathway which is cytochrome
    P450-dependent. This pathway is present in human bone marrow cells
    and peripheral blood neutrophils.

2. The formation of cytochrome P450 metabolite(s) was inhibited by
    SKF-525A, an inhibitor of cytochrome P450-dependent enzymes, but
    was unaffected in the presence of indomethacin or BW-755C.

3. Product(s) of the cytochrome P450-AA pathway were isolated, pooled
    and found to have a potent enhancing effect on human and murine
    bone marrow erythroid colony (CFU-E) growth.

REFERENCES

1. Lutton, J.D., R.D. Levere and N.G. Abraham. 1988. Biotechnology

of hemopoietic cells in culture: In: Biotechnology in Blood
Transfusion. Eds. C.Th.S. Sibinga, P.C. Das and L.R. Overby,
Kluwer Acad. Pubs., Boston, pp. 175-211.

2.   Levere, R.D., and N.G. Abraham. 1982. The bone marrow as a metabolic
organ. Am. J. Med. 73: 615-616.

3.   Abraham, N.G., J.D. Lutton, and R.D. Levere. 1985. Benzene modulation
of bone marrow hematopoietic and drug metabolizing systems.
Biochem. Arch. 1: 85-96.

4.   Lutton, J.D., K. Solangi, J.Y. Ran, J.D. Cooper, A. Goodman, N.G.
Abraham, and R.D. Levere. 1987. Development of a cytochrome
P-450 monooxygenase system in clonogenic hemopoietic cells.
Res. Comm. Chem. Path. Pharm. 56: 87-99.

5.   Needleman, P., J.Turk, B.A. Jakschik, A.R. Morrison, and J.B. Lefkowith.
1986. Arachidonic acid metabolism. Ann. Rev. Biochem. 55:
69-102.

6.   Schwartzman, M., M.A. Carroll, N.G. Abraham, N. Ferreri, E. Songu-Mize,
and J.C. McGiff. 1985. Renal arachidonic acid metabolism,
the third pathway. Hypertension 7 (SupplI): S136-S144.

7.   Schwartzman, M., K.L. Davis, J.C. McGiff, R.D. Levere, and
N.G. Abraham. 1988. Purification and characterization of
cytochrome P-450 dependent arachidonic acid epoxygenase from
human liver,. J. Biol. Chem. 263: 2536-2542.

8.   Schwartzman, M., N.R. Ferreri, M.A. Carroll, E. Songu-Mize, and
J.C. McGiff. 1985. Renal cytochrome P450 related arachidonate
metabolite inhibits ($Na^+ + K^+$) ATPase. Nature 314: 620-622.

9.   Escalante,B., W.C. Sassa, J.R. Falek, P. Yadagiri, and M.L. Schwartzman.
1989. Vasoreactivity of 20-hydroxyeicosatetraenoic acid is
dependent on metabolism by cyclooxygenase. J. Pharmacol. Exp.
Ther. 248: 229-232.

11.   Schwartzman, M.L., M. Balazy, J. Masferrer, N.G. Abraham, J.C.
McGiff, and R.C. Murphy. 1987. 12(R) - Hydroxyicosatetraenoic
acid: a cytochrome P450 dependent arachidonate metabolite that
inhibits Na^+K^+-ATPase in the cornea. Proc. Natl. Acad. Sci.
USA. 84: 8125-8129.

12.   Murphy, R.C., J.R. Falck, S. Lumin, P. Yadogiri, J.A. Zirrolli,
M. Balazy, J.L. Masferrer, N.G. Abraham, and M.L. Schwartzman.
1988. 12(R)-Hydroxyicosatri enoic acid: a vasodilator cytochrome
P-450 dependent arachidonate metabolite from bovine corneal
epithelium. J. Biol. Chem. 263: 17197-17202.

13.   Parnham, M.J., P.C. Bragt, and F.J. Zijlstra. 1981. Comparison
of the effects of inhibitors of cytochrome P-450 mediated reactions
on human platelet aggregation and arachidonic acid metabolism.
Biochem. Biophys. Acta 677: 165-173.

14.   Schwartzman, M.L., N.G. Abraham, J. Musferrer, M. Dunn, and J.C.
McGiff. 1985. Cytochrome P-450 dependent metabolism of arachi-
donic acid in bovine corneal epithelium. Biochem. Biophys
Res. Comm. 132: 343-351.

REGULATION OF HEME BIOSYNTHESIS IN CHICK EMBRYO LIVER CELLS

G.S. Marks, J.E. Mackie, S.A. McCluskey, and D.S. Riddick

Department of Pharmacology and Toxicology
Queen's University
Kingston, Ontario, Canada
K7L 3N6

ABSTRACT

According to current evidence heme controls the heme biosynthetic
pathway primarily by controlling translocation of inactive pre-ALA-S
from the cytosol into the mitochondrion, where ALA-S is active. A secondary
mechanism involves inhibition by heme of transcription of the ALA-S gene.
Porphyrinogenic drugs act by lowering a regulatory "free heme pool" by
three different mechanisms: (a) by mechanism-based inactivation of
cytochrome P-450 resulting in N-alkylprotoporphyrin formation and ferrochelatase inhibition, (b) by mechanism-based inactivation of cytochrome P-450
resulting in continuous heme destruction, (c) by enhanced generation
of active oxygen species which interact with an endogenous substrate
to form an inhibitor of uroporphyrinogen decarboxylase. It is also possible
that porphyrinogenic drugs may exert a direct effect on the nucleus to
increase formation of ALA-S mRNA.

The rate-controlling enzyme of the heme biosynthetic pathway is
δ-aminolevulinic acid synthase (ALA-S). This enzyme which is located
in the mitochondrion catalyzes the condensation of succinyl-CoA and glycine
to form δ-aminolevulinic acid (ALA). ALA passes out of the mitochondria
into the cytoplasm where two molecules condense together to form the
pyrrole, porphobilinogen (PBG). The enzyme involved in catalyzing this
reaction is δ-aminolevulinic acid dehydratase (ALA-D). PBG is converted
to a linear tetrapyrrole by the enzyme porphobilinogen deaminase. The
linear tetrapyrrole is transformed into uroporphyrinogen III (URO'GEN
III) by the enzyme URO'GEN III co-synthetase (Fig. 1). URO'GEN III is
then sequentially decarboxylated by the enzyme uroporphyrinogen decarboxylase
(UROG-D) to coproporphyrinogen III (COPRO'GEN III). in the process 7-carboxy,
6-carboxy, and 5-carboxy intermediates are formed. After passage into
the mitochondrion two of the propionic acid substituents of COPRO'GEN
are converted to vinyl groups, yielding protoporphyrinogen IX (PROTO'GEN
IX). In the next step of the pathway six hydrogen atoms are removed
from PROTO'GEN IX, with the formation of protoporphyrin IX (PROTO IX).
Ferrochelatase catalyzes the final step in the pathway, viz. the insertion
of ferrous iron into PROTO IX to form heme[1]. The heme is subsequently
incorporated into several hemoproteins with cytochrome P-450 synthesis
requiring more than half of the heme produced. Heme exerts feedback
repression on the synthesis of ALA-S. Normally this pathway is well
controlled and very little of the intermediate porphyrinogens accumulate.

Molecular Biology of Erythropoiesis
Edited by J. L. Ascensao *et al.*
Plenum Press, New York

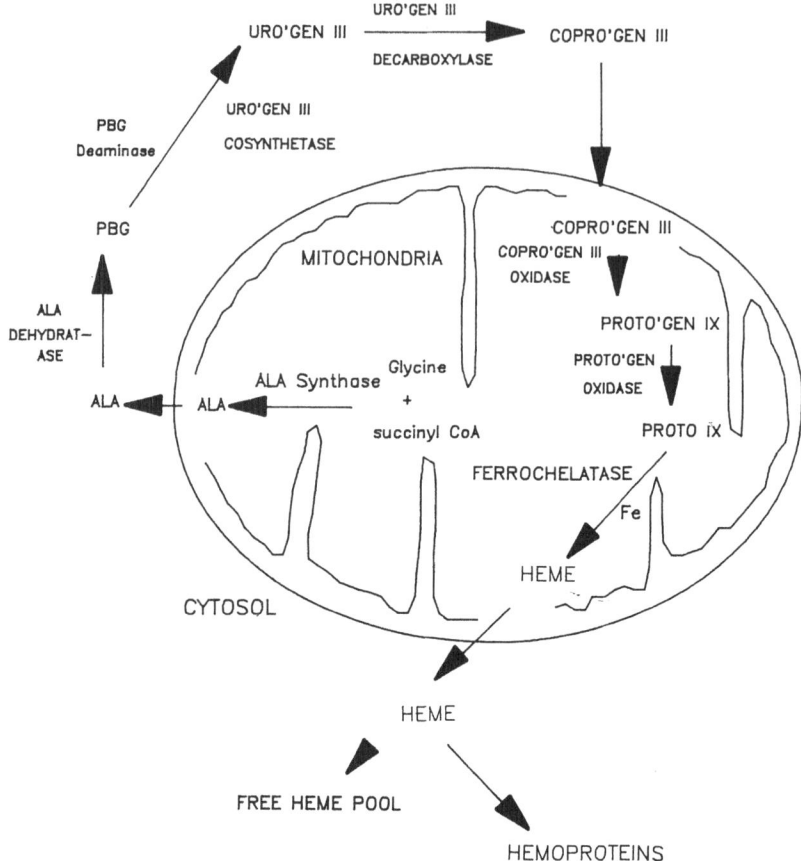

Fig. 1.   The heme biosynthetic pathway.

However, a variety of chemicals can interfere with the control of the pathway leading to the accumulation of porphyrinogen intermediates, which may in part undergo irreversible oxidation to porphyrins[1].

Granick[2] showed that the drug-induced synthesis of ALA-S in chick embryo hepatocyte culture was inhibited by heme. It was suggested that the repressor for ALA-S consisted of a protein aporepressor combined with heme, which functioned as a corepressor. Porphyrinogenic chemicals were hypothesized to displace heme from the aporepressor, thereby allowing ALA-S mRNA to be formed, increased synthesis of ALA-S to occur, and porphyrins to accumulate. The two major consequences which follow from this hypothesis are: 1) porphyrinogenic chemicals induce ALA-S activity by increasing the amount of ALA-S mRNA, and 2) heme prevents induction of ALA-S by repression of transcription.

Studies using actinomycin D and cycloheximide in chick embryo hepatocyte culture supported the hypothesis that the porphyrinogenic chemicals, allylisopropylacetamide (AIA, Fig. 2a) and 3,5-diethoxycarbonyl-1,4-dihydro-2,4,6-trimethylpyridine (DDC, Fig. 2b) exerted an effect at the transcriptional level by stimulating an increased synthesis of ALA-S mRNA[3]. The techniques utilized at the time were not sufficiently sensitive to confirm the increased synthesis of ALA-S mRNA. Recently the conclusions reached in these earlier studies have been confirmed using molecular biological techniques. Ades et al[4] devised a solution-hybridization assay for measuring steady-state

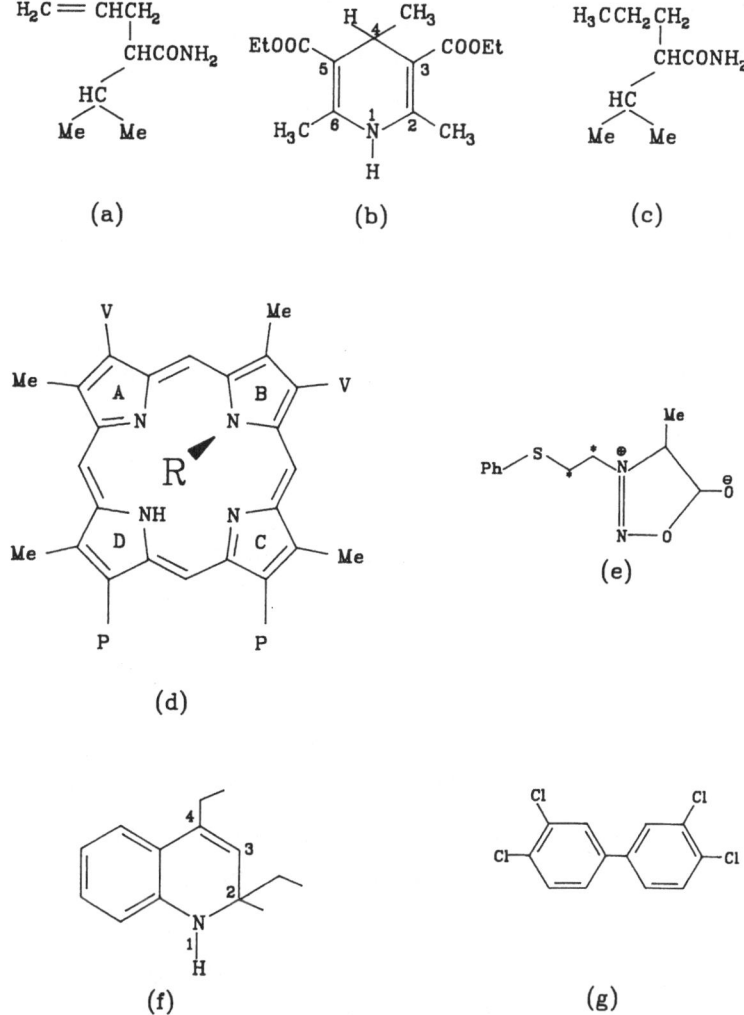

Fig. 2. (a) allylisopropylacetamide (AIA); (b) 3,5-die-
thoxycarbonyl-1,4-dihydro-2,4,6-trimethyl-
pyridine (DDC); (c) propylisopropylacetamide
(PIA); (d) N-methylprotoporphyrin IX (N-MePP,
R=Me); (e) 3-[2-(2,4,6-trimethylphenyl)-4-
methylsydnone (TTMS); (f) 2,4-diethyl-2-
dihydroquinoline (DMDQ); (g) 3,3',4,4'-
tetrachlorobiphenyl (TCBP).

levels of ALA-S mRNA which has been utilized by Hamilton et al[5] to demonstrate
an increase in steady-state levels of ALA-S mRNA in response to the porphyrinogenic chemicals, propylisopropylacetamide (PIA, Fig. 2c), DDC and glute-
thimide. It has been pointed out that the solution-hybridization assay
measures steady-state levels of mRNA rather than transcription rates.
However, there is at least preliminary evidence that the changes in steady-
-state ALA-S mRNA levels were the result of changes in transcription
rates of the ALA-S gene rather than a decrease in stability of the ALA-S
mRNA[5].

   The second consequence of Granick's hypothesis was that heme prevents
the induction of ALA-S by repression of transcription. In early studies

using actinomycin D and cycloheximide, it was shown that a post-transcriptional step in the formation of ALA-S was considerably more sensitive to inhibition by heme than was the transcriptional phase[3]. Thus the transcription of the ALA-S gene occured in the presence of concentrations of hemin which completely inhibited the post-transcriptional process. In subsequent studies[6] direct evidence was obtained for heme-mediated repression of transcription of the ALA-S gene; heme pretreatment significantly reduced the level of ALA-S mRNA.

An important additional mechanism whereby heme controls ALA-S activity is by inhibiting the transfer of an ALA-S precursor from the hepatic cytosol into mitochondria[7]. There is some evidence that heme exerts an effect on the translation of ALA-S mRNA into ALA-S[8]. Thus there is evidence that feedback repression of heme on ALA-S may be mediated by inhibition of transcription, inhibition of translation, and inhibition of translocation of ALA-S from the cytosol into the mitochondrion (Fig. 3).

Recent studies of Ades et al[4] and Hamilton et al[5] are compatible with the idea that heme regulates ALA-S primarily at the site of translocation of newly synthesized pre-ALA-S into the mitochondrion. Hamilton et al[5] used a combination of PIA and desferroxamine (an iron-chelator and inhibitor of heme synthesis) to induce increased levels of ALA-S activity and ALA-S mRNA in chick embryo hepatocytes. They were able

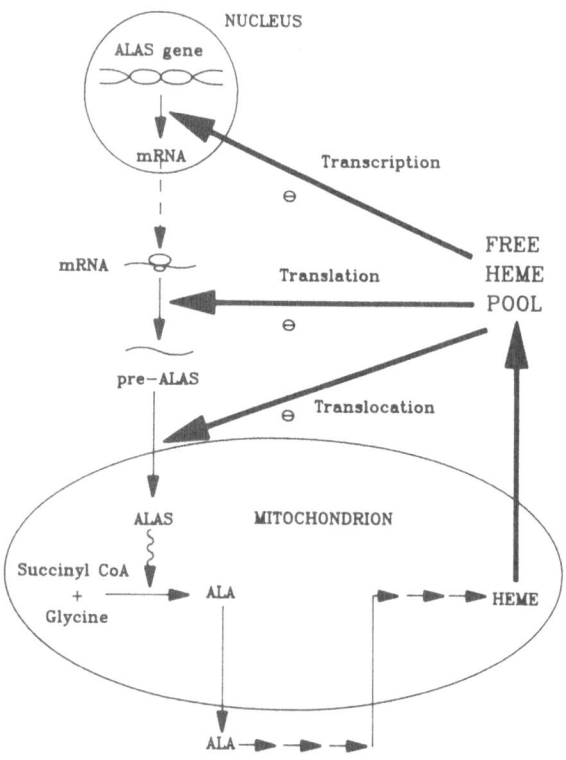

Fig. 3. Feedback repression of heme on ALA-S mediated by inhibition of transcription, inhibition of translation, and inhibition of translocation of pre-ALA-S from the cytosol into the mitochondrion. (From Marks, GS et al, FASEB J. 2: 2774, 1988, with permission)

to demonstrate that doses of heme which completely blocked induction of ALA-S activity, decreased induction of ALA-S mRNA by only 50%. Their results suggest that there is no direct correlation between induced levels of ALA-S activity and ALA-S mRNA. These investigators offer the following interpretation of their results: heme regulation of ALA-S is exerted primarily at the site of translocation of pre-ALA-S into mitochondria. Porphyrinogenic drugs act by an unknown mechanism at the transcriptional level to increase ALA-S mRNA. The resultant ALA-S activity is dependent upon the level of ALA-S mRNA and cellular free heme.

It has not yet been possible to measure "free heme"; "free heme" is thought to refer to either recently synthesized unbound heme, or heme just released from hemoproteins. An estimate of the "free heme" concentration has been made by Badawy et al[9] by measuring tryptophan pyrrolase activity. According to these investigators a regulatory free heme pool probably exists in the hepatic cytosol at a concentration range of 0.05-0.10 μM in close association with the hemoprotein tryptophan pyrrolase.

May et al[6] suggest the following sequence of events in drug-induced induction of ALA-S: drugs (e.g. phenobarbital) act at the level of the cytochrome P-450 gene causing increased formation of cytochrome P-450 mRNA leading to increased synthesis of the cytochrome P-450 apoprotein. Heme is diverted from the free heme pool for the synthesis of cytochrome P-450; the resulting lowering of the free heme pool leads to increased ALA-S mRNA formation and increased synthesis of ALA-S. Hamilton et al[5] point out that this model predicts that cytochrome P-450 mRNA should be induced before ALA-S mRNA. Using a solution-hybridization assay these investigators found that cytochrome P-450 mRNA and ALA-S mRNA were simultaneously induced by PIA in chick embryo hepatocyte culture. On the basis of this result and other experimental observations, the hypothesis of May et al[6] could not be supported.

In the remainder of this review we will seek to show that porphyrinogenic drugs act by lowering a regulatory "free heme pool". We will provide evidence that lowering the regulatory "free heme pool" by drugs can be brought about by three mechanisms:

(a) by mechanism-based inactivation of cytochrome P-450 resulting in N-alkylprotoporphyrin IX (N-alkylPP) formation and ferrochelatase inhibition.

(b) by mechanism-based inactivation of cytochrome P-450 resulting in continuous heme destruction.

(c) by enhanced generation of active oxygen species which interact with an endogenous substrate to form an inhibitor of UROG-D.

It is also possible that porphyrinogenic drugs may exert a direct effect on the nucleus to increase formation of ALA-S mRNA. Evidence for such a direct effect remains to be obtained.

The dihydropyridine DDC was shown to lower hepatic ferrochelatase activity when injected into rodents[10]. The dihydropyridine did not inhibit ferrochelatase when added to a crude enzyme preparation. These experimental findings were illuminated when N-methylprotoporphyrin IX (N-MePP) (Fig. 2d) was isolated from the livers of DDC-treated rodents and shown to be a potent inhibitor of ferrochelatase[11,12]. It was therefore clear why DDC itself did not inhibit ferrochelatase; only with the resulting accumulation of N-MePP did inhibition occur. The N-methyl group of N-MePP was found to arise from the 4-methyl group of DDC (Fig. 2b)[13-15]. Other porphyrinogenic analogues of DDC, viz., the 4-ethyl and 4-isobutyl analogues, when injected into phenobarbital-pretreated rats resulted in the accumulation of N-alkylPP's in which the N-alkyl substituent was derived from the 4-alkyl group of the DDC analogue in question[13,16].

The next question of interest is the origin of the protoporphyrin IX moiety of N-MePP. When DDC and related analogues are incubated with rat or chick embryo hepatic microsomes, mechanism-based inactivation of cytochrome P-450 is observed. This observation has given rise to the following hypothesis: DDC interacts with the active centre of cytochrome P-450 and the nitrogen atom of DDC undergoes one-electron oxidation (Fig. 4). The radical cation thus formed is unstable and a 4-methyl radical is ejected, which alkylates one of the four pyrrole nitrogens of the heme moiety of cytochrome P-450; loss of the iron atom from the alkylated heme moiety results in N-MePP formation[13,16].

It is of interest to determine to what extent the ALA-S inducing activity and porphyrinogenicity of 4-ethyl DDC could be attributed to the lowering of free heme levels by the inhibition of ferrochelatase caused by the N-ethylprotoporphyrin IX (N-EtPP) produced during metabolism of 4-ethyl DDC[17]. An additional means of manipulating the free heme pool by 4-ethyl DDC is the following: after cytochrome P-450 inactivation by heme alkylation, the alkylated heme moiety is released from the cytochrome P-450 apoprotein and the free apoprotein might combine with a heme molecule, thereby removing heme from the free heme pool. Support for this concept comes from work done with AIA[18]. In addition it is possible that DDC and analogues may act directly on the nucleus to increase levels of ALA-S mRNA[5]. In a recent study the contribution of ferrochelatase inhibition by N-EtPP to the induction of ALA-S activity caused by 4-ethyl DDC in chick embryo hepatocyte culture was investigated[17]. N-EtPP produced

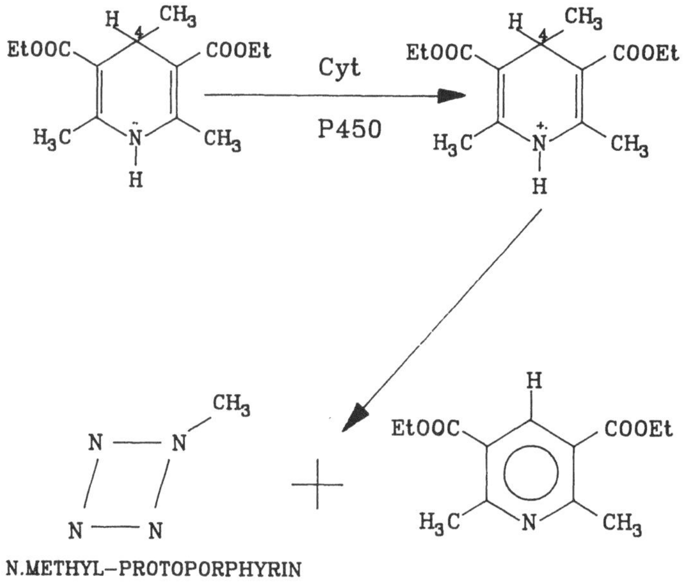

Fig. 4. Postulated mechanism of methyl transfer from DDC to the prosthetic heme of cytochrome P-450. Cytochrome P-450 is suggested to catalyze the formation of a radical cation from DDC; the active DDC intermediate aromatizes with transfer of the 4-methyl group to the heme moiety of cytochrome P-450.

an induction of ALA-S activity to 444% of control levels at 3 hours after administration (Fig. 5a). In contrast, 4-ethyl DDC caused an induction of ALA-S to 565% of control levels at 12 hours after administration (Fig. 5b). The delay in achieving maximal induction can be explained, at least in part, by the time required for formation of adequate levels of N-EtPP. The total induction, seen as the area under the ALA-S activity-time curve (Fig. 5b), caused by 4-ethyl DDC is greater than that caused by N-EtPP (1.9 times greater). It is apparent that the induction of ALA-S caused by 4-ethyl DDC cannot be accounted for solely by the inhibition of ferrochelatase caused by N-EtPP; another important mechanism(s) must be involved. These may be: (a) removal of heme from the free heme pool to reconstitute cytochrome P-450 following mechanism-based inactivation of cytochrome P-450 and/or (b) a direct effect of 4-ethyl DDC on the nucleus to increase ALA-S mRNA levels. In order to study the contribution of the destruction of cytochrome P-450 (and possible direct effects on the nucleus), the effect of 4-isobutyl DDC on ALA-S activity was determined. This compound causes the formation of N-isobutylprotoporphyrin (N-iBuPP) through the destruction of cytochrome P-450 heme, but ferrochelatase activity is not affected by this N-alkylPP. 4-Isobutyl DDC caused an increase in ALA-S activity to 289% of control levels at 3 hours after administration to chick embryo hepatocytes (Fig. 5c). This demonstrated that other mechanisms for elevating ALA-S activity other than ferrochelatase inhibition are operative with DDC analogues. It was anticipated that the combination of N-EtPP plus 4-isobutyl DDC would produce a curve similar to that seen with 4-ethyl DDC, since this represents a combination of an agent inhibiting ferrochelatase with an agent causing the destruction of cytochrome P-450 heme. The combination produced an induction of ALA-S activity considerably larger than was anticipated (Fig. 5d). At the 12 hour point, neither N-EtPP (Fig. 5a) nor 4-isobutyl DDC (Fig. 5c) caused any significant increase in the activity of ALA-S; however the combination produced over a 1200% increase in the activity of ALA-S[17]. There are several possible explanations for the synergistic effect of these two compounds on ALA-S activity. Inhibition of ferrochelatase by N-EtPP may reduce levels of free heme to the point where translocation of pre-ALA-S across the mitochondrial membrane and formation of active ALA-S becomes possible. 4-Isobutyl DDC may further lower heme levels by mechanism-based inactivation of cytochrome P-450 and in addition may exert a direct effect on the nucleus to increase levels of ALA-S mRNA. The increased level of ALA-S mRNA coupled with unimpeded translocation of pre-ALA-S across the mitochondrial membrane may explain the synergism. It should be possible to test this hypothesis by using the solution-hybridization assay to measure ALA-S mRNA and simultaneously estimating the "free heme pool" using tryptophan pyrrolase activity.

In recent years several chemicals have been found to share the mechanism by which dihydropyridines disrupt heme biosynthesis. The porphyrinogenic drug 3-[2-(2,4,6-trimethylphenyl)thioethyl]-4-methylsyndnone (TTMS) (Fig. 2e)[19,20] was found to cause mechanism-based inactivation of rat hepatic microsomal cytochrome P-450[21], ferrochelatase inhibition[20], and the hepatic accumulation of N-vinylprotoporphyrin IX (N-ViPP)[21]. The vinyl group in N-ViPP is thought to arise from the two carbon atoms in the thioethyl fragment of the 3-substituent of TTMS (marked with asterisks in Fig. 2e). The mechanism is believed to involve oxidation of the sydnone ring by cytochrome P-450 to an unstable ring that opens and breaks down liberating pyruvic acid and a diazo compound[21]. After loss of the nitrogen atoms from the diazo compound, the remaining portion of the molecule binds to the iron atom in the heme moiety of cytochrome P-450. After loss of the thiophenyl group as a leaving group and after the loss of iron, an N-vinyl group arises from the remainder of the TTMS molecule[21].

A third ring system which was found to cause mechanism-based inactiva-

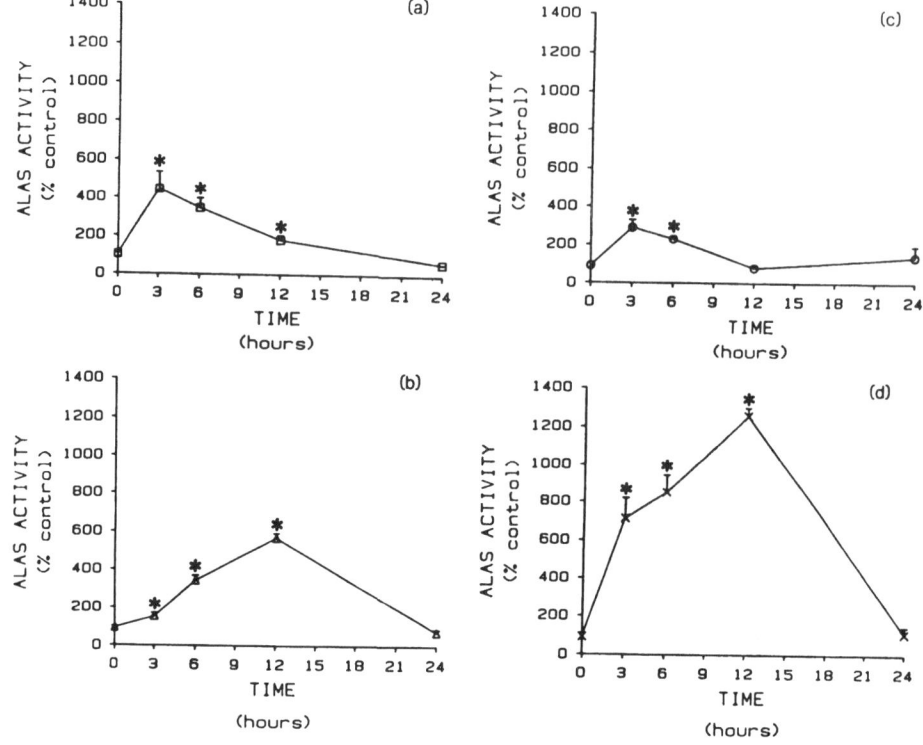

Fig. 5. Activity of δ-aminolevulinic acid synthase expressed as a percent of a 95% ethanol (0.064%) solvent control in chick embryo
liver cell culture at various times following the administration
of: (a) N-ethylPP; (b) 4-ethyl DDC; (c) 4-isobutyl DDC/ and
(d) 4-isobutyl DDC + N-ethylPP. All DDC compounds were given
at a concentration of 4 μM, and N-ethylPP was given at a 2 μM
concentration. Each point is the mean (± SD) of three determinations. Results were confirmed in an additional experiment
run with all compounds simultaneously. An asterisk (*) indicates
an increase over control levels at $P<0.05$. Control levels of
ALA-S activity were 216.5 ± 22.6 pmol/mg protein/30 min. (From
Mackie JE, Marks GS. Biochem Pharmacol. In press, with permission.)

tion of cytochrome P-450 with concomitant formation of N-MePP and N-EtPP
is the dihydroquinoline ring system (Fig. 2f). It is likely that the
source of the N-methyl and N-ethyl substituents of the N-alkylporphyrins
are the 2-ethyl and 2-methyl substituents of 2,4-diethyl-2-methyl-1,2-dihydro-
quinoline (DMDQ) liberated upon oxidation to the quinoline by cytochrome
P-450[22]. DMDQ was shown to cause ferrochelatase inhibition and protoporphyrin
IX accumulation in chick embryo hepatocyte culture[22].

    The structural features required for porphyrinogenicity in DDC,
TTMS, and DMDQ are: (a) a heterocyclic ring rendered unstable by oxidation
by cytochrome P-450 and (b) an appropriate ring substituent transferred
onto one of the four pyrrole nitrogen atoms of the heme moiety of cytochrome
P-450. One may now predict that any chemical that upon oxidation by
cytochrome P-450 might eject an alkyl radical, may result in mechanism-based
inactivation of cytochrome P-450, heme alkylation, ferrochelatase inhibition
and porphyrin accumulation[23].

    Allylisopropylacetamide (AIA) (Fig. la) is the prototype of a group
of compounds containing terminal olefine or acetylene groups, that lowers

hepatic free heme levels by the destruction of the heme moiety of cytochrome P-450[24,25]. In the process of mechanism-based inactivation of cytochrome P-450 by AIA, alkylation of a heme nitrogen occurs and after loss of iron from the alkylated heme, a 1:1 protoporphyrin-AIA covalently-bound adduct, devoid of ferrochelatase-inhibitory activity, is formed. The porphyrinogenicity of AIA is explained as follows: upon dissociation of the porphyrin-AIA adduct from cytochrome P-450 the apoprotein combines with fresh heme (presumably drawn from the free heme pool) to reconstitute cytochrome P-450[24,25]. A continuing interaction of cytochrome P-450 with AIA leads to marked depletion of free heme to the point where ALA-S is derepressed and porphyrins accumulate. The key structural feature in AIA for mechanism-based inactivation of cytochrome P-450 is the terminal olefinic group; ethylene itself is able to destroy cytochrome P-450 by prosthetic heme alkylation[25].

A variety of chemicals (e.g. 3,3',4,4'-tetrachlorobiphenyl (TCBP, Fig. 2g) cause the accumulation of uroporphyrin, heptacarboxylic acid porphyrin, and coproporphyrin in chick embryo hepatocyte culture. A modest inhibition of UROG-D by TCBP and other chemicals has been demonstrated in chick embryo hepatocyte culture, explaining, at least in part, the pattern of porphyrin accumulation[26]. TCBP and other chemicals are thought to induce and interact with a cytochrome P-450 isozyme resulting in generation of active oxygen species which interact with an endogenous hepatic substrate to form an inhibitor of UROG-D[27-29]. A second mechanism explains the large accumulation of porphyrins despite the modest reduction in UROG-D activity. Sinclair et al[30] have shown that a particular isozyme of cytochrome P-450, induced by polyhalogenated aromatic hydrocarbons such as TCBP, interacts with TCBP in the presence of NADPH to increase formation of an active oxidant that oxidizes URO'GEN and heptacarboxylic acid porphyrinogens to their respective porphyrins.

In summary porphyrinogenic drugs deplete a "regulatory free heme pool" by interacting with cytochrome P-450. Interaction with cytochrome P-450 results in either:

(a) N-alkylporphyrin formation and ferrochelatase inhibition,

(b) continuous heme destruction,

(c) enhanced generation of active oxygen species and UROG-D inhibition.

Porphyrinogenic drugs may act in an additional manner by a direct action on the cell nucleus to increase levels of ALA-S mRNA[5]. Heme controls ALA-S activity by: (a) inhibiting translocation of pre-ALA-S from the cytosol into the mitochondrion and (b) by inhibiting transcription of the ALA-S gene.

REFERENCES

1. Kappas, A., S. Sassa, and K.E. Anderson. 1982. The porphyrias, In Stanbury, J.B., Wyngaarden, J.B., Fredrickson, D.S. (eds): The metabolic basis of inherited disease. New York, McGraw-Hill, p. 1301.
2. Granick, S. 1966. The induction in vitro of the synthesis of δ-aminolevulinic acid synthetase in chemical porphyria: a response to certain drugs, sex hormones and foreign chemicals. J. Biol. Chem. 241: 1359.
3. Tyrrell, D.L.J., and G.S. Marks. 1972. Drug-induced porphyrin biosynthesis V. Effect of protohemin on the transcriptional

and post-transcriptional phases of δ-aminolevulinic acid synthetase induction. Biochem. Pharmacol. 21: 2077.

4. Ades, I.Z., T.M. Stevens, and P.D. Drew. 1987. Biogenesis of embryonic chick delta-aminolevulinate synthase: regulation of the level of mRNA by hemin. Arch. Biochem. Biophys. 253: 297.

5. Hamilton, J.W., W.J. Bement, P.R. Sinclair, J.F. Sinclair, and K.E. Wetterhahn. 1988. Expression of 5-aminolevulinate synthase and cytochrome P-450 mRNAs in chick embryo hepatocytes in vivo and in culture. Effect of porphyrinogenic drugs and haem. Biochem. J. 255: 267.

6. May, B.K., I.A. Borthwick, G. Srivastava, B.A. Pirola, and W.H. Elliott. 1986. Control of 5-aminolevulinate synthase in animals. Curr. Topics Cell. Regul. 28: 233.

7. Yamauchi, K., K. Hayashi, and G. Kikuchi. 1980. Translocation of δ-aminolevulinate synthase from the cytosol to the mitochondria and its regulation by hemin in the rat liver. J. Biol. Chem. 255: 1746.

8. Yamamoto, M., N. Hayashi, and G. Kikuchi. 1983. Translational inhibition by heme of the synthesis of hepatic δ-aminolevulinate synthase in a cell-free system. Biochem. Biophys. Res. Commun. 115: 225.

9. Badawy, A.A.B., C.J. Morgan, and N.R. Davis. 1985. Tryptophan pyrrolase and the regulation of mammalian heme biosynthesis, In: Nordmann Y. (ed): Porphyrins and Porphyrias. London, John Libbey Eurotext, INSERM, p. 69.

10. Onisawa, J., and R.F. Labbe. 1963. Effects of diethyl-1,4-dihydro-2, 4,6-trimethyl-pyridine-3,5-dicarboxylate on the metabolism of porphyrins and iron. J. Biol. Chem. 238: 724.

11. Tephly, T.R., A.H. Gibbs, and F. DeMatteis. 1979. Studies on the mechanism of experimental porphyria produced by 3,5-diethoxy-carbonyl-1,4-dihydrocollidine. Role of a porphyrin-like inhibitor of protohaem ferrolyase. Biochem. J. 180: 241.

12. Ortiz de Montellano, P.R., H.S. Beilan, and K.L. Kunze. 1981. N-methylprotoporphyrin IX: chemical synthesis and identification as the green pigment produced by 3,5-diethoxycarbonyl-1,4-dihydrocollidine. Proc. Natl. Acad. Sci. USA 78: 1490.

13. Ortiz de Montellano, P.R., H.S. Beilan, and K.L. Kunze. 1981. N-alkylprotoporphyrin IX formation in 3,5-diethoxycarbonyl-1,4-dihydrocollidine-treated rats. J. Biol. Chem. 256: 6708.

14. DeMatteis, F., A.H. Gibbs, P.B. Farmer, and J.H. Lamb. 1981. Liver production of N-alkylated porphyrins caused in mice by treatment with substituted dihydropyridines. FEBS Lett. 129: 328.

15. Tephly, T.R., B.L. Coffman, G. Ingall, G.S. Abou Zeit-Har, H.D. Tabba, and K.M. Smith. 1981. Identification of N-methylprotoporphyrin IX in livers of untreated mice and mice treated with 3,5-die-thoxycarbonyl-1,4-dihydrocollidine: source of the methyl group. Arch. Biochem. Biophys. 212: 120.

16. Augusto, O., H.S. Beilan, and P.R. Ortiz de Montellano. 1982. The catalytic mechanism of cytochrome P-450. Spin-trapping evidence for one-electron substrate oxidation. J. Biol. Chem. 257: 11288.

17. Mackie, J.E., and G.S. Marks. Synergistic induction of δ-amino-levulinic acid synthase activity by N-ethylprotoporphyrin IX and 3,5-diethoxycarbonyl-1,4-dihydro-2,6-dimethyl-4-isobutyl-pyridine. Biochem. Pharmacol. (In press).

18. Farrel, G.C., and M.A. Correia. 1980. Structural and functional reconstitution of hepatic cytochrome P-450 in vivo. Reversal of allylisopropylacetamide-mediated destruction of the hemoprotein by exogenous heme. J. Biol. Chem. 255: 10128.

19. Stejskal, R., M. Itabashi, J. Stanek, and Z. Hruban. 1975. Experi-

mental porphyria induced by 3-[2-(2,4,6-trimethylphenyl)-thioethyl]-4-methylsydnone. Virchows Arch. B. Cell. Path. 18: 83.

20. Sutherland, E.P., G.S. Marks, L.A. Grab, and P.R. Ortiz de Montellano. 1986. Porphyrinogenic activity and ferrochelatase-inhibitory activity of sydnones in chick embryo liver cells. FEBS Lett. 197: 17.

21. Ortiz de Montellano, P.R., and L.A. Grab. 1986. Inactivation of cytochrome P-450 during catalytic oxidation of a 3-[(arylthio) ethyl]sydnone: N-vinyl heme formation via insertion into the Fe-N bond. J. Am. Chem. Soc. 108: 5584.

22. Lukton, D., J.E. Mackie, J.E. Lee, G.S. Marks, and P.R. Ortiz de Montellano. 1988. 2,2-Dialkyl-1,2-dihydroquinolines: cytochrome P-450 catalyzed N-alkylporphyrin formation, ferrochelatase inhibition, and induction of 5-aminolevulinic acid synthase activity. Chem. Res. Toxicol. 1: 208.

23. Marks, G.S., S.A. McCluskey, J.E. Mackie, D.S. Riddick, and C.A. James. 1988. Disruption of hepatic heme biosynthesis after interaction of xenobiotics with cytochrome P-450. FASEB J. 2: 2774.

24. DeMatteis, F. 1978. Loss of liver cytochrome P-450 caused by chemicals. Damage to the apoprotein and degradation of the heme moiety, In: DeMatteis, F., Aldridge, W.N. (eds): Handbook of experimental pharmacology, Vol. 44. Berlin. Springer-Verlag, p. 95.

25. Ortis de Montellano, P.R., and M.A. Correia. 1983. Suicidal destruction of cytochrome P-450 during oxidative drug metabolism. Ann. Rev. Pharmacol. Toxicol. 23: 481.

26. Lyon, M.E., J.A. Owen, and G.S. Marks. 1988. Xenobiotic mediated inhibition of hepatic uroporphyrinogen decarboxylase activity in 17-day-old chick embryo liver cells in culture. Biochem. Pharmacol. 37: 1123.

27. Billi, S.C., W. De Calmanovich, and L.C. San Martin de Viale. 1986. Rat liver porphyrinogen carboxylase inhibitor as a function of the degree of hexachlorobenzene-induced porphyria, In: Morris, C.R., Cabral, J.R.D. (eds): Hexachlorobenzene: proceedings of an international symposium, Vol. 77. Lyon, IARC Scientific Publications, p. 487.

28. Cantoni, L., D. Dal Fiume, H. Rizzardini, and R. Ruggieri. 1984. In vitro inhibitory effect on porphyrinogen carboxylase of liver extracts from TCDD treated mice. Toxicol. Lett. 20: 211.

29. Smith, A.G., and J.E. Francis. 1987. Chemically-induced formation of an inhibitor of hepatic uroporphyrinogen decarboxylase in inbred mice with iron overload. Biochem. J. 246: 221.

30. Sinclair, P., R. Lambrecht, and J. Sinclair. 1987. Evidence for a cytochrome P-450-mediated oxidation of uroporphyrinogen by cell-free extracts from chick embryos treated with 3-methylcholanthrene. Biochem. Biophys. Res. Commun. 146: 1324.

LONG-TERM EXPRESSION OF THE HUMAN β-GLOBIN GENE AFTER RETROVIRAL

TRANSFER INTO PLURIPOTENT HEMATOPOIETIC STEM CELLS OF THE MOUSE

Richard E. Gelinas*, Michael A. Bender,
A. Dusty Miller, and Ulrike Novak

Department of Molecular Medicine
Fred Hutchinson Cancer Research Center
1124 Columbia Street
Seattle, WA 98104

ABSTRACT

We have studied the regulation of the human β-globin gene after
retroviral transfer into a variety of transformed and normal hematopoietic
cells. After transfer into murine erythroleukemia cells (MEL) expression
from the human β-globin gene responds to inducers of erythroid maturation
in parallel to the endogenous murine globin genes. After infection of
human BFU-E, RNA expression from the virally-transferred β-globin gene
was measured at 2.5%-5% of the endogenous β-globin level. The most improved
globin vectors can transfer the human β-globin gene into pluripotent
hematopoietic stem cells in mouse bone marrow. Mice reconstituted with
infected marrow show human β-globin RNA and protein expression in peripheral
blood cells for over 4 months. In these animals, both myeloid and lymphoid
cells carry the integrated provirus at a level of about 1 copy per cell.
In serial transplantation experiments, bone marrow from these animals
is capable of repopulating secondary and tertiary recipient animals which
go on to show long-term human β-globin expression. Retroviral vectors
thus provide a practical way to refine models of globin gene regulation
through in vivo tests and to evaluate the feasibility of protocols for
gene addition therapy.

INTRODUCTION

Gene addition therapy for chronic anemias such as β-thalassemia
major is based on the ability to introduce a β-globin gene and necessary
regulatory sequences into pluripotent stem cells of the bone marrow. In
theory, a cure could result if normal or near-normal levels of β-globin
protein derived from the transferred gene resulted in adequate levels
of HbA tetramer formation in erythroid progeny cells. Successful therapy
would also require that such globin expression be confined to cells of
the erythroid lineage, and that the integration of the provirus did not
disturb the expression of host genes. Retroviruses have several properties

Molecular Biology of Erythropoiesis
Edited by J. L. Ascensao et al.
Plenum Press, New York

which make them attractive as gene transfer vectors (Miller et al., 1989), and vectors have been described which can transfer marker genes into pluripotent hematopoietic stem cells (PHSCs) of the mouse (Dick et al., 1985; Keller et al., 1985; Lemischka et al., 1986). We are studying whether retroviral vectors which can transfer globin genes can form the basis of a gene addition therapy for thalassemia.

We have designed retroviral vectors which include the human β-globin gene along with the neomycin phosphotransferase gene (neo) as a dominant selectable marker and used them to study the regulation of the transferred human β-globin gene in murine erythroleukemia (MEL) cells and human BFU-E (Bender et al., 1988). In these vectors, the human β-globin gene is carried in a reverse transcriptional orientation compared to the transcription from the long terminal repeat (LTR) of the vector. Similar vectors have been described by several other groups (Cone et al., 1987; Karlsson et al., 1988; Lerner et al., 1987). In our experiments with MEL cells, the transferred β-globin gene was regulated along with the endogenous globin genes and produced RNA at a substantial fraction of the endogenous level. We found that the β-globin virus could be improved after DNA sequences which interfered with viral replication were removed (Miller et al., 1988). The improved β-globin vector can transfer the β-globin gene into the PHSCs of mice, which means that long-term in vivo studies on the expression and regulation of the virally-transferred gene are now possible (Bender et al., 1989).

METHODS

The details of the construction of replication-defective retroviral vectors for the human β-globin gene, the generation and characterization of amphotropic and ecotropic virus-producing cell lines from plasmid forms of the vectors, and the infection and selection of MEL cells, human BFU-E, and mouse bone marrow cells have all been described (Bender et al., 1988; Miller et al., 1988). Methods for monitoring hematopoietic reconstitution of mice with virus-infected bone marrow, preparation of DNA and RNA from blood, bone marrow, spleen, and thymus, and determination of proviral copy number by Southern hybridization have been described (Bender et al., 1989).

The structures of RNA probes and their use to quantitate mRNA transcribed from the virally-transferred human β-globin gene or the mouse β^{dmajor} β^s-globin genes by solution hybridization followed by ribonuclease digestion and gel electrophoresis of the protected fragments have also been described (Bender et al., 1988). Human β-globin mRNA derived from the virally transferred gene could be quantitated in the presence of mRNA from the endogenous copies of this gene because of a 6 bp insertion in the 5' untranslated region of the former gene as described (Bender et al., 1988). Human β-globin polypeptide was detected in MEL cells and in mouse peripheral blood smears with an immunofluorescence assay based on a monoclonal antibody which is specific for human β-globin chains under the conditions used (Bender et al., 1988).

In experiments designed to test the effect of sequences derived from the region 5' to the ε-globin gene on β-globin expression, linearized plasmids were introduced into MEL cells by electroporation (Forrester et al., 1989). MEL cells (1-2 x 10^7/ml) were electroporated in HEPES-buffered glucose (Chu et al., 1987) with the apparatus described by (Bradshaw et al. 1987) equipped with a 240 μF capacitor. The electric field was 1000 volts/cm. After electroporation, the cells were cultured for 15-24 hours before being plated in semisolid medium with G418 for the isolation of individual colonies.

RESULTS

The LNβ*HP vectors and human β-globin expression in MEL cells

Among the first β-globin vectors we prepared were two in which the marked human β-globin gene or an intronless minigene were cloned into a replication-defective derivative of the Moloney murine leukemia virus (Bender et al., 1988). Human globin sequences extended from an HpaI site in the β-globin promotor (-815) to the first PstI site 3' to the gene (+2161). The globin fragment was cloned downstream of the neo gene such that the globin gene would be transcribed from its own promoter and opposite to transcription from the long terminal repeat of the retrovirus. Maps for these vectors (LNβ*HP or LNβ*HP MG) are presented in Fig. 1. G418-resistant clonal cell lines which contained unrearranged proviruses were identified by Southern hybridization and virus titers were determined. The highest titers for the amphotropic packaging cell lines with an unrearranged provirus are given in Table 1. The LNβ*HP MG virus, which lacked both globin introns, was produced at a 40-fold higher titer than the LNβ*HP virus. Both viruses transmitted the appropriate provirus without rearrangement after infection of MEL cells, but globin introns proved to be essential for RNA expression.

Populations of MEL cells were infected by exposure to medium containing the LNβ*HP or LNβ*HP MG retroviruses, selected in G418, and induced to differentiate with hexamethylene bisacetamide (HMBA). Steady state levels of human β-globin, mouse $β^{maj}$-globin, or mouse $β^{min}$-globin RNA were measured by RNA protection analysis. In pools of infected MEL cells, human β-globin expression was 10% of the endogenous mouse $β^{maj}$-globin RNA level as summarized in Table 1. RNA expression from the intronless minigene could not be detected. Although the mouse $β^{maj}$-globin RNA levels increased 55 to 190 times during the 4-day induction period, expression from the human β-globin gene increased between 1.5 to 7 times in isolated clones. The reduced level of induction noted for the virally transferred globin gene

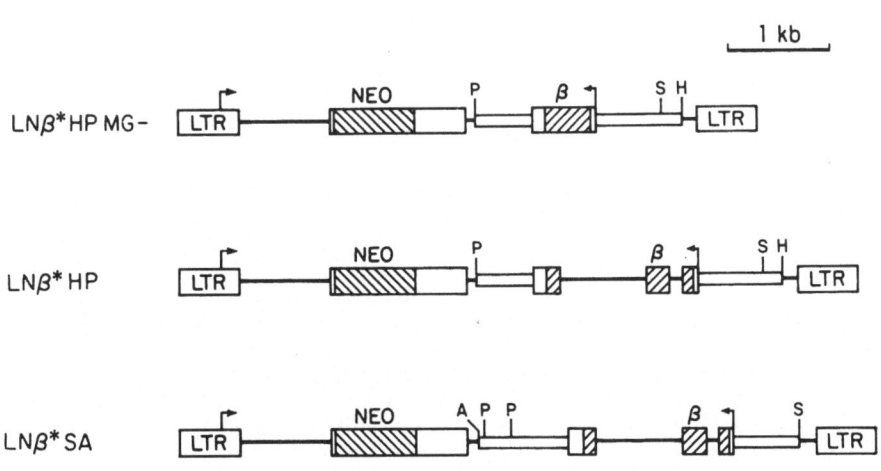

Fig. 1. Maps of the β-globin retroviruses. Arrows denote origins of transcription. β, LTR, and NEO stand for the human β-globin gene (or minigene), the retroviral long terminal repeat, and the neomycin phosphotransferase gene respectively. P, S, H, and A mark the cleavage sites for PstI, SphI, HpaI, and AvrII within the globin gene; other restriction cleavage sites for these enzymes are not shown.

Table 1.   Characteristics of LNβ*HP and LNβ*HPMG Retroviruses (Bender et al., 1988

	Maximum Titer[a]	Helper Virus[b]	RNA Expression in MEL Cells[c]
	CFU/ml	FFU/ml	hβ/mβ^{maj}
LNβ*HP	5×10^4	<2	0.10
LNβ*HP MG	2×10^6	<2	<0.001

a.   The highest titer on NIH3T3 TK$^-$ cells, from the clones with the correct structure (Bender et al., 1988).

b.   As measured on virus-producing clones with the highest titer with the S$^+$L$^-$ assay (FFU, Focus-forming units).

c.   Ratio of human β-globin RNA to murine β^{maj}-globin RNA after 6 days of HMBA treatment.

was due in part to a low level of expression prior to induction. Although this 'preinduced' expression was comparable in many experiments to preinduced expression of the mouse β^{min}-globin we also noted in these cells, it may have been caused in part to a position effect arising from having selected the cells for neo expression. RNA transcribed from the human β-globin gene was active in directing protein synthesis. Immunofluorescence analysis of clonal MEL lines infected with the LNβ*HP virus using a mono-clonal antibody specific for human β-globin chains showed that 30 to 100% of the cells fluoresced brightly after 4 days of induction, with little fluorescence prior to induction.

The construction of an improved β-globin vector LNβ*SA

We explored the basis for the large difference in titers observed between the LNβ*HP or LNβ*HP MG viruses by examining the structures of the intracellular RNAs transcribed from these proviruses in the virus--producing cell lines as well as from the retroviral particles themselves. Northern analysis showed that in contrast to the simple pattern of RNAs observed for the parental neo-vector with no globin gene, the cell line which produced the LNβ*HP retrovirus contained multiple viral RNAs, of which only a minor proportion were full-length transcripts of the integrated provirus (Miller et al., 1988). In addition, subgenomic length RNA molecules were the prodominant species recovered from the virions produced from these cells. The proportion of full length RNA was much higher in virions and producing cells for the LNβ*HP MG virus, which helped account for its higher titer.

The map locations of the various subgenomic sized RNAs detected in the LNβ*HP producing cells suggested that sequences within the second intron as well as the promoter of the globin gene were leading to termination of transcription of the integrated provirus. The termination could be mediated by AAUAAA sequences in the primary transcript leading to poly(A) addition (Miller et al., 1988). Since only full-length RNA is capable of transmitting the genome of the LNβ*HP virus, the globin promoter sequences which were associated with RNA termination were deleted from the next virus which was constructed, LNβ*SA. In this virus, the β-globin promoter was cut back to the SphI site at -615 and 3' flanking sequence up to the AvrII site at +2482 was addedin order to include the region shown to influence expression of the human γ-globin gene in transgenic mice (Behringer et al., 1987; Trudel et al, 1987). LNβ*SA also made use of an improved vector which was designed to have a reduced probability of producing viral gag proteins or to give rise to replication-competent helper virus (Fig. 1). The titer of the LNβ*SA virus was 2×10^5 cfu/ml which was higher than the LNβ*HP virus, but was still inferior to the minigene virus.

Introns are necessary for expression of the human β-globin gene

We studied the question of how the globin introns might influence expression by preparing derivatives of the LNβ*SA virus in which the β-globin gene contained either the first (LNβ*SA1+2-) or the second intron (LNβ*SA1-2+). A minigene version was prepared as well, and the set of four viruses were used to infect MEL cells. After selection in G418, individual clones were pooled, induced with HMBA, and human and mouse β-globin RNA levels were measured. The data in Table 2 shows that the presence of the globin second intron had a negative correlation with virus-titer but a positive correlation with steady state RNA levels. The LNβ*SA virus (both introns) and the LNβ*SA1-2+ virus (second intron only) produced similar levels of correctly initiated human β-globin RNA after induction. In contrast, human β-globin RNA was reduced 10-fold in cells infected with the LNβ*SA1+2- (first intron only), and human β-globin RNA could not be detected in cells infected with the minigene virus. Steady state β-globin RNA levels showed a similar pattern in the virus-producing PA317 cells lines (Miller et al., 1988) and in transgenic mice (Brinster et al., 1988; R. Gelinas, unpublished results) suggesting that determinants associated with the second intron but not the first are required for efficient β-globin expression.

Expression of the marked human β-globin gene in BFU-E colonies

The LNβ*SA virus proved to be more efficient at transferring the

Table 2. Characteristics of LNβ*SA Set of Retroviruses

	Maximum Titer CFU/ml[a]	RNA Expression in MEL Cells[b]
LNβ*SA	2×10^5	100
LNβ*SA 1-2+	5×10^5	50
LNβ*SA 1+2-	2×10^6	4
LNβ*SA MG	4×10^6	<0.3

a. The highest titer on NIH3T3 TK⁻ cells, from the PA317 clone with the correct structure (Miller et al., 1988)

b. Average steady state human β-globin RNA level in virus-infected pools of MEL cells after 6 days of HMBA treatment, expressed as a percentage of the LNβ*SA level.

β-globin gene to BFU-E than the LNβ*HP virus. Bone marrow from normal donors was enriched for mononuclear cells, infected by cocultivation for 24 hours with irradiated virus-producing fibroblasts, and cultured in semi-solid medium with and without G418 for 14 days under conditions which allowed growth of BFU-E erythroid progenitor cells (Bender et al., 1988). The 6 base pair insertion in the 5' untranslated region of the virally transferred β-globin gene rendered its transcripts quantifiable in the presence of β-globin RNA derived from the endogenous globin genes. Data in Table 3 compares the BFU-E infection efficiency and RNA expression levels demonstrated by the two viruses. Although the LNβ*SA virus infected BFU-E more readily than the LNβ*HP virus, both viruses directed similar levels of steady state RNA in the nucleated erythroid cells which comprised the bursts, in spite of the fact that the LNβ*SA virus included additional 3' sequences previously shown to have enhancer-like properties in transgenic mice. Globin expression from the LNβ*SA provirus was confined to the erythroid lineage, since human β-globin RNA could not be detected in CFU-GM progeny cells which also grew in this assay.

Human β-globin expression in mice reconstituted with infected bone marrow

We next tested whether the LNβ*SA virus would result in efficient infection of murine PHSCs, and if so, what level of β-globin expression would be observed and how stringently this expression would be regulated (Bender et al., 1989). In these long-term reconsitution and expression studies, we used anemic W/W^v mice as recipients of virus infected bone marrow. One practical consequence of the W/W^v genotype is that marrow ablation was not necessary to permit hematopoietic reconstitution with donor bone marrow as would be required for normal mice. In some experiments bone marrow donors were used (C57BL/6J) that had a different hemoglobin haplotype in comparison to the W/W^v mice, which simplified determination of the time course of hematopoietic reconstitution. Bone marrow donor animals were treated with 5-fluorouracil 2 to 5 days before marrow was collected from leg bones. After infection by cocultivation with virus-- producing fibroblasts for 1 to 5 days, marrow in most experiments was selected in G418 for an additional 2 days before being infused into the tail veins of recipient animals. The marrow from 2 to 8 donor animals was infused into each mouse which received G418-selected marrow, and

Table 3.   Tissue Specific β-globin Expression After Viral Transfer to Bone Marrow Progenitor Cells.

Virus	Maximum Titer[1] CFU/ml	BFU-E Infection[2] Efficiency	hβ*/hβwt RNA expression in progeny of	
			BFU-E	CFU-G/M
LNβ*HP	5 x 10^4	0.01%	5%	not measured
LNβ*SA	2 x 10^5	0.2-1.0%	0.9-2.5%	not detected

1.   From amphotropic PA317-derived virus producing cell lines, as measured on NIH3T3 TK$^-$ cells.

2.   Ratio of G418-resistant to total erythroid colonies.

when no preselection was used, infected marrow from a single donor was given to recipient animals. No helper virus was detected by the XC assay in any mice which received LNβ*SA-infected bone marrow. Human β-globin expression was detected and monitored in mice reconstituted with virus-infected marrow by immunofluorescence analysis on peripheral blood smears.

In one experiment (M8 in Bender et al., 1989), marrow from C57BL donor mice was infected by cocultivation for 1 day before selection in G418. Three out of four animals successfully engrafted with the donor marrow as shown by greater than 75% conversion to the donor hemoglobin type by 4 months post-transplantation. Hemoglobin haplotypes were monitored by using cellulose acetate gel electrophoresis (Fig. 2, lower panel). Two of the three animals which engrafted in this experiment expressed human β-globin protein in 40-50% of their circulating erythrocytes through-out this time as well (Fig. 2, upper panel). In other experiments in which bone marrow donor animals were normal and coisogenic (+/+) to the W/W^V recipients, all 4 mice transplanted with infected selected marrow showed long term β-globin expression by immunofluorescence. Overall, three-fourths of mice which reconstituted with infected bone marrow showed long-term human β-globin expression (Bender et al., 1989). In parallel experiments, only 2 of 12 mice reconstituted with infected marrow that had not been selected showed expression of human β-globin polypeptide for over four months. In these animals, only 2% of the peripheral blood

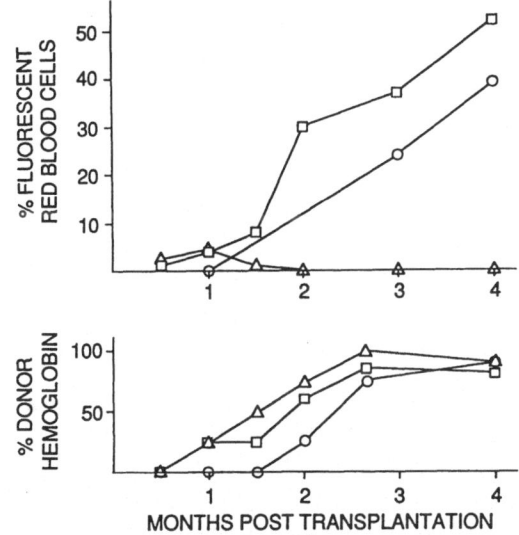

Fig. 2. Hematopoietic reconstitution of human β-globin expression in mice after gene transfer into bone marrow. The top panel shows the percentage of circulat-ing erythrocytes that expressed detectable human β-globin as a function of time after transplantation, as measured by immunofluorescence. The bottom panel shows the percentage of donor mouse hemoglobin in erythro-cytes as a function of time after trans-plantation, as determined by electro-phoretic separation of mouse hemoglobin on cellulose acetate.

cells were positive for human β-globin expression.

In several experiments, animals which received selected marrow showed
human β-globin expression by immunofluorescence of peripheral blood between
2 and 6 weeks post-transplantation which disappeared completely past
10 weeks.   In two animals which showed this transient expression, but
which successfully engrafted with donor marrow as shown by conversion
to the donor hemoglobin type, we explored the basis for the loss of expression.   No provirus was detected by Southern analysis of DNA isolated
from hematopoietic organs (data not shown) which suggested that the PHSCs
which repopulated these animals were simply not infected and that the
transient human β-globin expression derived from infected progenitor
cells with limited repopulation potential.   In no case was evidence obtained
for an inactive provirus in reconsitituted mice which showed no human
β-globin expression.

The tissue specificity of expression of the virally-transferred
human β-globin gene was analyzed by the RNase protection assay on RNA
from blood, marrow, spleen and thymus as shown in Fig. 3.   Mouse β^S-globin
RNA represented 6 to 12% of the total poly(A)+ RNA in peripheral blood,

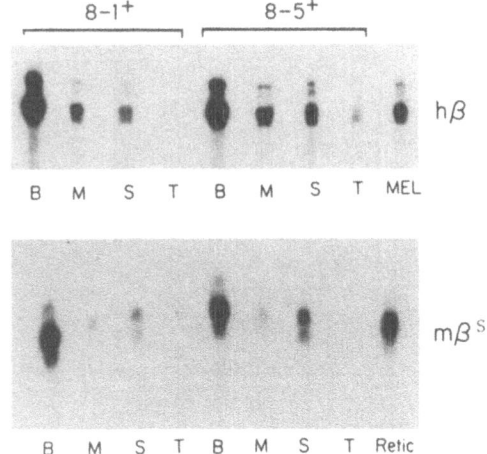

Fig. 3.   RNA protection analysis of human
β-globin RNA in mouse hematopoietic
tissues.   RNA was isolated from the
blood (B), bone marrow (M), spleen
(S), and thymus (T) of two of the
mice analyzed in Fig. 2.   4µg of
RNA was analyzed for the presence
of human β-globin (hβ) as shown in
the upper panel, and 80 ng of RNA
was analyzed for mouse β^S-globin ($m\beta^S$)
mRNA in the lower panel using a ribonuclease protection assay.   MEL indicates RNA from a MEL cell line
expressing a virally transduced human
β-globin gene.   Retic indicates a
sample of mouse reticulocyte total
nucleic acid.   The structures of
the RNA probes and the sizes of the
protected fragments have been described (Bender et al., 1988).

while human β-globin RNA was expressed in blood at 0.24 to 1.5% of the
mouse globin level in experiments in which G418-selected marrow was used
(Table 4).   In mice transplanted with nonselected marrow, human β-globin
varied from 0.01 to 0.04% of the mouse level, consistent with the lower
infection rate observed in these animals (see below).   Although the absolute
amounts of human β-globin and mouse β^S-globin RNA were lower in bone
marrow and spleen than in blood cells, the ratio of human β-globin to
mouse β^S-globin RNA in these tissues was similar to their ratio in blood,
as expected since spleen and bone marrow are normal sites of erythropoiesis.
Thymus samples from a few mice showed a human to mouse globin RNA ratio
greater than the ratio observed in blood, consistent with ectopic human
β-globin expression in the thymus at an extremely low level (mouse 8-1[+]
in Table 4).

To determine the proportion of hematopoietic cells containing a
provirus, and to determine if cells derived from all hematopoietic lineages
were infected with the LNβ*SA virus, DNA was isolated from bone marrow,
spleen, and thymus of animals which had expressed human β-globin 4-5
months after transplantation.   The marrow and spleen of all animals transplanted with selected marrow contained in LNβ*SA provirus at about one
copy per hematopoietic cell (Bender et al., 1989 and Fig. 4, last lane).
Proviral copy number varied from <0.03 to 2 copies per cell in thymic
tissue, as has been noted previously for the variable degree of reconsitition
of lymphoid tissues with W-anemic mice (Harrison et al., 1976; Harrison
et al., 1979).   The presence of the LNβ*SA provirus in both myeloid and
lymphoid tissues suggested that the full hematopoietic system and been
reconstituted with donor marrow, not just its erythroid element.   The
pattern and intensity of the different provirus integrations suggested

Table 4.   RNA Expression in Mice 4-5 Months After Transplantation

Mouse	Tissue	Copy#	$m\beta^S$	$h\beta$	Ratio(%)
8 - 1[+]	Blood	N.E.	11.0	0.16	1.5
	Marrow	++++	0.48	0.0098	2.0
	Spleen	++++	1.0	0.0082	0.82
	Thymus	0	0.015	0.00038	2.5
8 - 5[+]	Blood	N.E.	12.0	0.18	1.5
	Marrow	++++	0.84	0.014	1.7
	Spleen	++++	2.0	0.025	1.3
	Thymus	++++	0.019	0.0025	13.0
13 - 1[+]	Blood	N.E.	6.8	0.025	0.37
	Marrow	++++	2.0	0.0090	0.45
	Spleen	++++	1.6	0.0099	0.62
	Thymus	++++	0.059	0.00085	1.44
13 - 5[+]	Blood	N.E.	6.8	0.028	0.41
	Marrow	++++	1.3	0.011	0.85
	Spleen	++++	1.7	0.0062	0.36
	Thymus	++++	0.077	0.00036	0.47

Proviral copy number and globin RNA expression were measured in long-term expressing mice,
4-5 months after transplantation at the time of sacrifice. Proviral copy number was determined
by Southern blotting: ++++, at least 1 copy/cell. $m\beta^S$ and $h\beta$-globin RNA values are given as
the percentage of poly(A)[+] RNA as determined by RNA protection analysis. Ratio: the ratio
of human β- to mouse β^S-globin RNA expressed as a percentage. Mouse numbers with a + sign
received G418-selected marrow; N.E., not evaluated.

that in most cases one or a few infected stem cells contributed to hematopoietic tissue regeneration (Bender et al., 1989). In contrast to the
high copy number observed in mice which received selected marrow, mice
reconstituted with non-selected marrow had about 0.05 copies of the provirus
per cell (data not shown).

To test the proliferative capacity of the marrow in β-globin positive
primary marrow recipients, after 4-5 months the animals were sacrificed
and marrow was infused into secondary W/W^v recipients. All (3 out of 3)
secondary recipients transplanted with marrow from one of the mice described
in Fig. 2 converted to donor hemoglobin type and showed ≥50% immunofluor-
escese-positive erythrocytes. Four months post-transplantation, one
mouse was sacrificed and found to be reconstituted with cells containing
the same viral integration of the LNβ*SA provirus as had been observed
in the primary recipient (Fig. 4). Another portion of the bone marrow
from the secondary recipient mouse was infused into two additional mice.
After seven months,' these tertiary recipient animals also showed 32 to
46% immunofluorescent-positive erythrocytes. Thus, the LNβ*SA virus
originally infected a multipotential cell with a high proliferative capacity,
or PHSC.

Normal levels of β-globin expression with hypersensitive site constructions

The human β-globin RNA expression levels observed after viral transfer
into BFU-E or into murine PHSC were only about 1-2% of the endogenous
levels, which is below the level which would have therapeutic value in
β^o-thalassemia patients. In an attempt to raise β-globin expression
after gene transfer, we tested whether the portion of the β-like globin
gene cluster upstream of the ε-globin gene that had previously been shown
to confer high level expression on a β-globin gene in transgenic mice
(Grosveld et al., 1987) could be reduced in size while retaining functional
activity. This region of the DNA had also been shown to contain hypersensitive sites which were conserved in erythroid cell lineages (Forrester
et al., 1987). Four strong hypersensitive sites were identified in these
studies in erythroid tissues which map at 6.1, 10.9, 14.7, and 18 kb
5' to the ε-globin gene. For one test, a 23 kb region 5' of the ε-globin
gene implicated by the transgenic mouse experiments was reduced in size
to 8 kb by removal of sequences between the hypersensitive sites and
cloned adjacent to a 3 kb human β-globin reporter gene (Forrester et
al., 1989). The construct was linearized and introduced into MEL cells
by electroporation. The neo gene was included as a selectable marker
in trans. In most clones, the amount of human β-globin RNA equaled the
mouse β^{maj}-globin RNA after induction with HMBA or dimethyl sulfoxide,
and correction of the data for the copy number of the human β-globin
gene. An equal level of human and mouse β-globin RNA was also observed
with a related construction in which the DNA region containing the hypersensitive sites was reduced still further, to about 2.5 kb in size.
Retroviruses which include this region are being investigated.

DISCUSSION

A retrovirus has been developed which can transfer the human β-globin
gene into PHSCs of the mouse. In some experiments, all transplanted
mice which reconstituted with donor marrow expressed the human β-globin
protein for more than 4 months, and up to 50% of peripheral blood cells
contained detectable polypeptide. Overall, 6 out of 8 mice which reconstituted after receiving infected, selected marrow showed long-term expression
of human β-globin RNA and protein. Previous attempts to use retroviral
vectors to transduce the human β-globin gene into murine bone marrow
have demonstrated the difficulty in obtaining animals which showed persistent

144

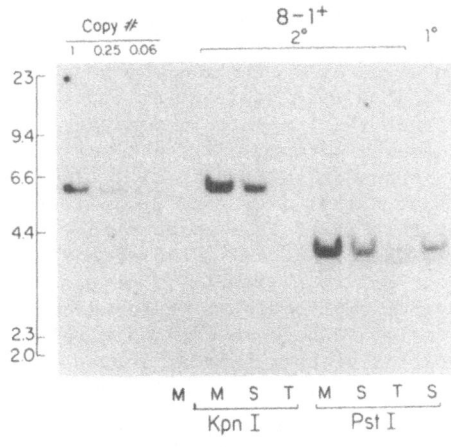

Fig. 4. DNA analysis of hematopoietic tissues from a secondary bone marrow recipient. Four months after transplantation, a secondary bone marrow recipient was sacrificed and DNA was isolated from bone marrow (M), spleen (S), and thymus (T), and 5 μg samples were digested with either KpnI (to determine proviral copy number) or PstI (to reveal proviral integrations), and analyzed by Southern blotting with the human β-globin mini-gene probe. DNA samples from the mouse which received marrow from mouse 8-1[+] are marked 2°. Spleen DNA from mouse 8-1[+] was processed in parallel (lane marked 1°). The sizes of molecular weight standards are given at the left in kilobase pairs. For more details, see Bender et al., 1989.

expression of the transduced gene in a high proportion of hematopoietic cells (Dzierzak et al., 1988; Karlsson et al., 1988). The titer of the vector used in our studies, LNβ*SA, was 20 times higher than the vectors used by (Dzierzak et al. 1988) who also observed long-term β-globin expression in a minority (8 out of 104) of reconstituted mice. In another study in which W/W[V] recipient mice were employed (Karlsson et al., 1988) no long-term expressing mice were obtained. Although virus titer probably contributed to the success of our experiments, other differences in the infection protocols employed preclude a definitive conclusion.

Large amounts of infected and selected marrow were required to assure successful engraftment in our experiments, which suggest that the reconstitutive potential of this marrow was decreased. Future experiments will attempt to minimize the loss of repopulation ability of the infected marrow. In several experiments, spleen colonies (12-day CFU-S) were analyzed for the presence of the LNβ*SA provirus and human β-globin RNA and proved to be positive. Unfortunately, the ability to efficiently infect CFU-S did not always correlate with the ability of the same sample of infected marrow to reconstitute W/W[V] mice and give rise to long-term expression. In these experiments, therefore, the conditions necessary to infect CFU-S and to infect and preserve the repopulation potential of PHSC after selection are apparently different (Bender et al., 1989).

Several configurations of the human β-globin gene were evaluated in retroviruses in order to identify one which would show regulated expression after transfer to erythroid cells and which could also be produced at high titers. A human β-globin minigene or intronless gene met the latter requirement but not the former. In contrast, other intronless genes such as the rat growth hormone mini gene were expressed at a high level after retroviral transfer (Miller et al., 1984). Comparison of human β-globin genes with either intron or no introns revealed that the second intron, but not the first, is necessary for expression of normal levels of steady state RNA. Ironically, the second intron, as well as a region upstream of the β-globin promoter, contained sequences which interfered with retrovirus replication. We found that the interfering region near the promoter could be removed, and viral titers improved as a result. The resulting vector, LNβ*SA, showed an improved ability to infect human BFU-E, and proved capable of infecting PHSC in mice.

Although the murine transplant model makes use of inbred animals, it remains a useful animal model for evaluating new β-globin transducing viruses and infection protocols. Surprising, expression in normal erythroid cells such as mouse peripheral blood cells or human BFU-E progency was about one-tenth the level observed in MEL cells, which reinforces the need for in vivo tests of retroviral vectors. The LNβ*HP and LNβ*SA viruses produced the same level of RNA in BFU-E progeny, in spite of the fact that the latter virus included additional 3' sequences in region of the putative enhancer. This may suggest that the 3' region of the human β-globin gene may not act as a classical enhancer at all, or that it may mediate other aspects of globin regulation.

The next challenges for β-globin virus development will be to elevate β-globin expression to a level above the therapeutic threshold and to improve the efficiency of stem cell infection. The next generation of β-globin retroviral vectors will include DNA from the 'locus activation' or 'dominant control region which was recently identified upstream of the human ε-globin gene (Grosveld et al., 1987; Forrester et al., 1987). Our preliminary experiments show that this region can be decreased in size substantially while biological activity is retained. Infection efficiencies can be improved by increasing retroviral titers and by purification of the target cell population. It should be possible to increase titers of β-globin transducing viruses still further by identifying and eliminating sequences which interfere with viral replication.

ACKNOWLEDGEMENTS

We thank Edith Harris and Wendy Magis for help with the experiments, and Jenny Torgerson for preparation of the manuscript. M.A.B. was supported by the Medical Scientist Training Program of the National Institutes of Health. These studies were supported by grants from the NIH (HL-37073, CA-30924, HL-36444, and AM-31232).

REFERENCES

Behringer R.R., R.E. Hammer, R.L. Brinster, R.D. Palmiter, and T.M. Townes. 1987. Two 3' sequences direct erythroid-specific expression of human β-globin genes in trangenic mice. Proc. Natl. Acad. Sci. USA 84: 7056-7060.
Bender, M.A., A.D. Miller and R.E. Gelinas. 1988. Expression of the human β-globin gene after retroviral transfer into murine erythroleukemia cells and human BFU-E cells. Mol. Cell Biol. 8: 1725-1735

Bender, M.A., R.E. Gelinas, and A.D. Miller. 1989. A majority of
    mice show long-term expression of a human β-globin gene after retroviral
    transfer into hematopoietic stem cells. <u>Mol. Cell Biol.</u> 9: 1426-1434.
Bradshaw, H.D., W.W. Parson, M. Sheffer, P.J. Lioubin, E.R. Mulvihill,
    and M.P. Gordon. 1987. Design, construction, and use of an electro-
    porator for plant protoplasts and animal cells. <u>Anal. Biochem.</u>
    166: 342-348.
Brinster, R.L., J.M. Allen, R.R. Behringer, R.E. Gelinas, and R.D. Palmiter.
    1988. Introns increase transcriptional efficiency in transgenic
    mice. <u>Proc. Natl Acad. Sci. USA</u> 85: 836-840.
Chu, G., H. Hayakawa, and P. Berg. 1987. Electroporation for the
    efficient transfection of mammalian cells with DNA. <u>Nuc. Acids
    Res.</u> 15: 1311-1326.
Cone, R.D., A.W. Benarous, D. Baorto, and R.C. Mulligan. 1987. Regulated
    expression of a complete human β-globin gene encoded by a transmissible
    retrovirus vector. <u>Mol. Cell Biol.</u> 7: 887-897.
Dick, J.E., M.C. Magli, D. Huszar, R.A. Phillips, and A. Bernstein.
    1985. Introduction of a selectable gene into primitive stem cells
    capable of long-term reconstitution of the hematopoietic system
    of W/W^v mice. <u>Cell</u> 42: 71-79.
Dzierzak, E.A., T. Papayannopoulou, and R.C. Mulligan. 1988. Lineage-
    specific expression of a human β-globin gene in murine bone marrow
    transplant recipient reconstituted with retrovirus-transduced stem
    cells. <u>Nature</u> 331: 35-41.
Forrester, W.C., S. Takegawa, T. Papayannopoulou, G. Stamatoyannopoulos,
    and M. Groudine. 1987. Evidence for a locus activation region:
    the formation of developmentally stable hypersensitive sites in
    globin-expressing hybrids. <u>Nuc. Acids Res.</u> 15: 10159-10177.
Forrester, W.C., U. Novak, R. Genlinas, and M. Groudine. 1939. Molecu-
    lar analysis of the human β-globin locus activation region. <u>Proc.
    Natl. Acad. Sci. USA</u> 86: (in press).
Grosveld, F., G.B. van Assendelft, D.R. Greaves, and G. Kollias.
    1987. Position-independent, high level expression of the human
    β-globin gene in transgenic mice. <u>Cell</u> 51: 975-985.
Harrison, D.E., and C.M. Astle. 1976. Population of lymphoid tissues
    in cured W-anemic mice by donor cells. <u>Transplantation</u> 22: 42-46.
Harrison, D.E., C.M. Astle, and J.A. DeLaittre. 1979. Processing by
    the thymus is not required for cells that cure and populate W/W^v
    recipients. <u>Blood</u> 54: 1152-1157.
Karlsson, S., T. Papayannopoulou, S.G. Schweiger, G. Stamatoyannopoulos,
    and A.W. Nienhuis. 1987. Retroviral-mediated transfer of genomic
    globin genes leads to regulated production of RNA and protein. <u>Proc.
    Natl. Acad. Sci. USA</u> 84: 2411-2415.
Karlsson, S., D.M. Bodine, L. Perry, T. Papayannopoulou, and A.W. Nienhuis.
    1988. Expression of the human β-globin gene following retroviral-
    mediated transfer into multipotential hematopoietic progenitors
    of mice. <u>Proc. Natl. Acad. Sci. USA</u> 85: 6062-6066.
Keller, G., P. Paige, E. Gilboa, and E.F. Wagner. 1985. Expression
    of a foreign gene in myeloid and lymphoid cells derived from multipotent
    haematopoietic precursors. <u>Nature</u> 318: 149-154.
Lemischka, I.R., D.H. Raulet, and R.C. Mulligan. 1986. Developmental
    potential and dynamic behavior of hematopoietic stem cells. <u>Cell</u>
    45: 917-927.
Lerner, N., S. Brigham, S. Goff, and A. Bank. 1987. Human β-globin
    expression after gene transfer using retroviral vectors. <u>DNA</u> 6:
    573-582.
Miller, A.D. 1989. Retroviral Vectors. In: Current Topics in Microbiology
    and Immunology. (N. Muzyczka, Ed.) (In Press).
Miller, A.D., E.S. Ong, M.G. Rosenfeld, I.M. Verma, and R.M. Evans.
    1984. Infectious and selectable retrovirus containing an inducible
    rat growth hormone minigene. <u>Science</u> 225: 993-998.

Miller, A.D., and C. Buttimore. 1986. Redesign of retrovirus packaging
    cell lines to avoid recombination leading to helper virus production.
    Mol. Cell Biol. 6: 2895-2902.
Miller, A.D., M.A. Bender, E.A.S. Harris, M. Kaleko, and R.E. Gelinas.
    1988. Design of retroviral vectors for transfer and expression
    of the human β-globin gene. J. Virol. 62: 4337-4345.
Trudel, M., and F. Costantini. 1987. A 3' enhancer contributes to the
    stage-specific expression of the human β-globin gene. Genes Dev.
    1: 954-961.

RETROVIRAL GENE TRANSFER IN MICE: THE USE OF A

UNIQUE PACKAGING LINE IMPROVES EFFICIENCY

Charles S. Hesdorffer, Dina Markowitz,
Maureen Ward, Norma B. Lerner, and Arthur Bank

Department of Genetics and Development and the
Division of Hematology
Columbia University,
New York, NY 10032

The principles of gene therapy, although superficially simple, involve three important prerequisites: (1) GENE TRANSFER, (2) GENE EX- PRESSION, and (3) SAFETY.

GENE TRANSFER

First, the new gene must be inserted into the appropriate target cell and must remain there. Retroviral vectors are used to insert foreign genes into cells. Retroviruses have the unique ability to enter cells with high efficiency, and to integrate their genome into the cells' chromo- somes[1-4]. Retrovirus mediated gene transfer has been used successfully in hematopoietic stem cells in vitro. Infected cells have subsequently been followed by detecting viral DNA integrants, and their fate determined following transplantation into lethally irradiated mice as described by Lemischka and his colleagues[5].

Two specific issues remain regarding the use of retroviruses as vectors of the DNA to be inserted into recipient cells, such as the bone marrow stem cells of mice, primates or humans. These issues are efficiency and safety. Forty-eight hour incubation of bone marrow with the producer line and subsequent 24 hours of selection with high doses of G418 followed by transplantation of these cells into lethally irradiated mice show that only 20% of hematopoietic stem cells become infected with the exogenous gene[5].

Secondly, what are the dangers associated with the use of retroviruses? We must be aware of the possibility that replication-competent viruses may form, and that the proliferation of those viruses will lead to multiple random integrations in the host genome. These integrations could potential- ly result in the activation of harmful genes such as oncogenes[6,7] with the disastrous consequence of possible tumor formation. To avoid this complication, packaging lines have been mutated such that by themselves they are not transmissable, replication-competent viruses[2,8-11]. To date the most commonly used packaging line has a deleted psi sequence, which is the signal necessary for packaging viral RNA. The required gag, pol, and env genes of this retrovirus are intact. Theoretically,

Molecular Biology of Erythropoiesis
Edited by J. L. Ascensao *et al.*
Plenum Press, New York

there should be no formation of wild type "helper" virus by this packaging line. However, when this packaging line is transfected with replication-defective retroviral vectors containing the psi sequence necessary for their own packaging, wild type virus has been demonstrated, presumably as a result of a single recombination event[12,13]. Miller and Buttimore attempted to circumvent this problem by creating additional mutations. These mutations included deletions in the 3'LTR of the helper virus component, or additional deletions of portions of the 5'LTR. One of these defective amphotropic constructs (where further deletions of the 5'LTR are employed) has been used to produce a retroviral packaging line, PA317[13]. It is apparent that two recombinational events could still produce intact retrovirus. Attempts at producing lines containing both the 3' and 5'LTR deletions as well as the packaging mutation have failed because of the low titers obtained with these constructs.

We have been involved in the development of a retrovirus in which three mutational events are combined to produce a packaging line which is both safe and efficient[14].

Using the Moloney murine leukemia virus (Mo-MULV) present in the plasmid 3PO, which has a 134 base pair deletion of the psi packaging sequence, as starting material, two plasmid constructs were made: 1) Pgag-polgpt, containing the gag and pol genes of Mo-MULV; and 2) penv, containing the env gene from this virus. To generate cell lines expressing gag-pol and env regions from the different plasmids, 3T3 cells were cotransfected with gag-polgpt and env DNA's. Recipient cells were selected with media containing mycophenolic acid and the GP+E clones isolated were tested for the production of reverse transciptase(RT). The high RT-producing clones were analyzed for env protein expression by metabolic labelling followed by immunoprecipitation with env antiserum. One of these clones of GP+E which was found to be high RT and medium-high env expressing (GP+E86) was transfected with the ΔNEO retroviral vector. The titers of the GP+E86+ΔNEO clones ranged from 10^2 to 1.7×10^6 CFU/ml while the titers obtained using the plasmid 3PO in a packaging line in a control experiment were consistently lower and ranged from 8×10^2 to 6.5×10^5 CFU/ml. Additional experiments were performed in which GP+E86 cells were tested for their ability to package the retroviral vector N2. Seventeen of the 22 (77%) clones generated N2 viral titers of $>10^5$ CFU/ml. On the other hand, 5 of 11 (45%) of clones of 3PO cells, similarly transfected with N2, generated titers of $>10^5$ CFU/ml[14]. Thus, the in vitro efficiency of this new retroviral packaging lines (GP+E86) was demonstrated.

The efficiency of this packaging line can be further improved by increasing the titer of virus production. It has been demonstrated that by first transfecting the vector into an amphotropic packaging line and then using the supernatant to infect an ecotropic line, a log higher viral titer can be obtained[15]. Experiments using GP+E86 infected with N2 have confirmed this interesting finding. In future, this method of producer line preparation may become universal.

A number of in vitro studies were performed to demonstrate the safety of this unique packaging line[14]. First, supernatants from GP+E86+ΔNEO clones were used to infect 3T3 cells. G418 resistant clones were selected and allowed to develop into a confluent layer. Supernatants from these plates were used to infect fresh 3T3 cells. These 3T3 cells were again selected for G418 resistance and none was found, neither was any RT detected in the supernatants obtained from these secondarily infected 3T3 cells. This indicates the lack of viral rescue of ΔNEO from the initial 3T3 cells infected. In a more stringent test for recombinant infectious retrovirus, 3T3 cells were again infected with supernatants from high

titer GP+E86+ΔNEO clones. These 3T3 cells were then passaged continuously
for 1 month without G418 selection. Supernatants were tested for ΔNEO
production by infecting fresh 3T3 cells and testing for G418 resistance.
As in the previous experiment, no G418-resistant cells were noted. A
final test was performed to detect a transfer of packaging function.
3T3 cells were electroporated with N2 DNA and pools of G418 resistant
clones were collected. Supernatant from GP+E86 cells were used to infect
these 3T3-N2 pools. If the 3T3-N2 pools became infected with wild type
virus secreted from the GP+E86 cells, indicating a transfer of packaging
function, these cells would begin to secrete N2 virus. Supernatant from
the infected 3T3-N2 pools was harvested and used to infect further 3T3
cells which were then assayed for the presence of N2 virus by exposure
to G418. No G418-resistant clones were detected indicating that GP+E86
cells are unable to transfer packaging function.

GENE EXPRESSION

In order to cure thalassemia it is not enough to merely introduce
a new β-Globin gene into a marrow stem cell but this gene must function
appropriately. Cloned β-Globin genes have been introduced into mouse
erythroleukemia (MEL) cells which can be induced to differentiate in
culture. When the human β-Globin gene in introduced into MEL cells,
this gene is coregulated with the endogenous mouse globin genes[16]. Furthermore, these exogenous genes, removed from their normal chromosomal contexts
and introduced into MEL cells, continue to be appropriately regulated[16-19].
These studies indicate that the β-Globin gene fragments used in these
experiments include the sequences required in cis for specific transcriptional activation of β-Globin genes during erythroid differentiation.

Other studies of β-Globin gene expression have involved the introduction
of cloned globin genes into the mouse germ line by microinjection of
DNA and the analysis of gene expression during development of the transgenic
mouse[20-30]. These experiments have shown that cloned adult β-Globin
genes are expressed in a tissue specific as well as a stage specific
manner after transfer into the mouse germ line[20,21,23]. From these studies
the structural requirements for genetic constructs expressed in mouse
gene transfer experiments were determined.

In our initial studies we used constructs containing the human β-Globin
gene and the neomycin resistance gene to transfect psi 2 cells and obtain
retrovirus infecting 3T3 cells and MEL cells[24]. These studies demonstrated
the importance of the orientation of the β-Globin gene. Only clones
in which the β-Globin gene was inserted in the opposite orientation to
the retrovirus were found to have intact β-Globin genes by southern blot
analysis.

Further studies have shown that specific gene sequences, known as
enhancers, are required for gene expression. There are at least three
regions within and surrounding the β-Globin gene that are important for
its expression[25-27]: 1) Sequences 5' to the β-Globin gene including a
TATA box, CAAT region and a GC-rich region; 2) regions within the β-Globin
gene itself; and 3) enhancer sequences 3' to the gene, most specifically
that described by Trudel and Costantini which is believed to end 900bp
3' to the β-Globin gene itself[28]. More recently, Grosveld and his co-workers
have demonstrated a number of hypersensitive sites 5' to the Epsilon
globin gene and 3' to the β-Globin gene which are important in position--
independant high level expression of the β-Globin gene in transgenic
mice[29]. Presumably, these DNA sequences must be inserted with the new
genome into the hematopoietic stem cells to achieve successful gene therapy.

Having evaluated the theories upon which gene therapy is based, the effect on the results of in vivo studies, using mice as the animal model, will be domonstrated. The impact of the unique retroviral packaging line GP+E86 on the efficiency of gene transfer is discussed. In order to properly evaluate the results of hematopoietic marrow transplantation, recipient mice must receive a dose of radiation which will kill all mice which do not successfully reconstitute, while not harming the mice themselves due to radiation toxicity. Our studies with C57BL/6J mice demonstrated that the dose of radiation which fulfills these requirements is 1100 rads. (Figure 1).

Bone marrow transplantation was performed using the standard method as described by Dick et al[3]. The packaging line GP+E86 was transfected with the vector N2. In the first set of transplant experiments the titer of virus obtained was 10^6 CFU/ml. Mice given varying amounts of radiation were injected with 1×10^6 nucleated marrow cells. Following transplantation these mice were kept in sterile cages, on sterile bedding and were maintained on autoclavable mouse food and water supplemented with Tetracycline (4gm/L). Table 1 describes the results of southern blot analyses performed on mice sacrificed at various intervals to determine the success of gene transfer. 62% efficiency of gene transfer is better than that previously seen using a vector containing the neomycin resistance gene alone. (Figures 2 and 3 are examples of southern blot analyses performed on these mice to determine the presence of the neomycin resistance gene and the degree of chimerism created following transplantation with marrow incubated with GP+E86+N2.)

Utilizing the transfect-infect method, higher viral producing titers have been obtained. A further 18 C57BL/J6 mice underwent bone marrow transplantation with cells incubated with GP+E86 producing N2 at a titer of >10^7 CFU/ml.(Table 2.) Thus, the overall success of gene transfer in these two groups of experiments was 65%.

In vivo safety testing of the packaging line is complex because C57BL/J6 mice express endogenous retrovirus from the age of 3 months. Evaluation of serum for RT may thus be misleading. No RT has been found in the blood of transplanted or control mice. A more stringent test involves the incubation of serum with 3T3 cells for 1 month followed by selection with G418. In this system only 3T3 cells infected with a retrovirus carrying the neomycin resistance gene will survive. Supernatant obtained from these clones can be tested for the presence of RT activity. No G418 resistant clones should be found in the control mouse serum tested. However, it is possible that mice transplanted with GP+E86+N2 can demonstrate G418 resistant 3T3 cells though no wild type retrovirus related to the packaging line was created. Theoretically, the endogenous mouse retrovirus can package N2. Thus, G418 resistant clones may be demonstrated and RT may be found in the supernatant produced by these infected cells; a false positive test. Two hundred days after transplantation, G418 resistant 3T3 cells with RT in the supernatant was found in one mouse. It is likely that this finding reflects the incorporation of N2 by endogenous mouse retrovirus. All ongoing tissue culture studies of GP+E86 have not demonstrated wild type retrovirus formation. Thus the demonstration of wild type virus formation in vivo in one mouse is probably a false positive result. However, all transplanted mice will continue to be evaluated in this manner.

We have begun studies aimed at transplanting mice with GP+E86 containing a vector with a human β-Globin gene oriented in the reverse direction.

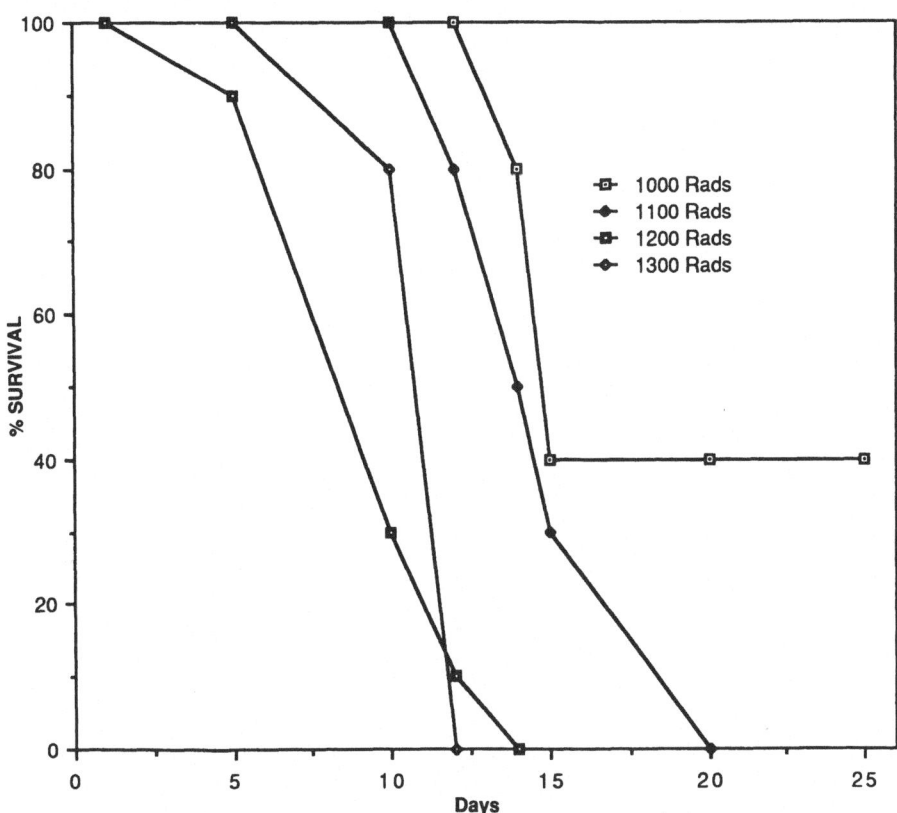

Fig. 1. A radiation killing curve was constructed by exposing
groups of ten C57B1/J6 female mice to increasing doses
of radiation to determine the optimal dose of radiation
which would kill all untransplanted mice without any
deaths due to radiation toxicity. The mice were subjected to radiation at a dose rate of 120rads/min using
a cesium source. The total radiation dose was given in
two divided doses in one day. 2/3 followed by 1/3, 3-4
hours later. The appropriate dose which resulted in a
50% death rate by Day 12-15, a 100% death rate by Day
21, and no deaths prior to Day 10, was 1100rads.

The construct we have created contains a 3.9kb β-Globin gene cloned
into PLNL at the single Sall site. Using the plasmid BpBsv, the β-Globin
fragment extends from the Sphl site approximately 400 bp 5' to the β-Globin
gene, to the Sphl site 1.8kb 3' to the β-Globin gene including the putative
β-Globin gene enhancer recently reported by Trudel and Costantini[28].
Using the transfect-infect method to create the virus producing packaging
line, titers of greater than 10^4 CFU/ml have been obtained. <u>In vitro</u>
evidence suggests that the entire β-Globin gene is present in infected
3T3 cells. Studies will be done using live mice analagous to ôur previous
experiments with the N2 vector. The far lower titers of GP+E86+PLNL-B

Table 1. Results of Bone Marrow Transplantation With Neomycin Resistance Gene-Titer $1*10^5$

NUMBER OF MICE SACRIFICED BY DAY SACRIFICED

RADS	NUMBER	DAY 14	DAY 90	DAY 200	RAD DEATH	LATE DEATH	ALIVE on day 290
900	16	1					
950	12	2	2	2	1	3	2
1000	10	2	2	2	2	0	2
Mice tested		5	4	4			
Percent of mice with gene		40%	75%	75%			

SUCCESSFUL GENE TRANSFER - 8/13 (62%)

Data from southern blot analysis, using Neo probe, of spleen DNA from 13 mice following transplantation with marrow incubated with GP+E86 packaging N2 at a titer of 10 particles/ml. Only WEHI conditioned medium was used during the 72 hour incubation period. Successful transplantation is noted in 62% of the mice examined.

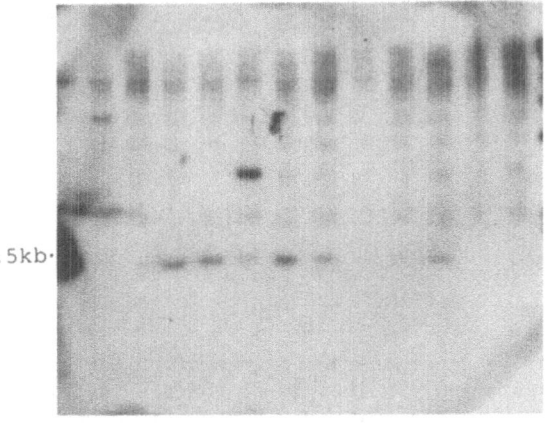

Fig. 2. DNA was extracted from spleen and marrow
cells by standard techniques[30]. 10micro
gm of DNA were digested with ECO RI and
hybridised with an N2 probe. Southern
Blot analysis demonstrated the presence
of the neomycin resistance gene, the 1.5kb
fragment, in Lanes 4,5,6,7,8,10,11.

Lane 1: Probe (N2 cut with ECO RI)
Lane 2: Negative control-Radiation
         only mouse, no transplantation
Lane 3: Negative control-Untransplanted
         mouse
Lane 4: 950 Rad, spleen DNA, 14 days post
         transplant-mouse 1
Lane 5: 1000 Rad, spleen DNA, 14 days post
         transplant-mouse 2
Lane 6: 950 Rad, spleen DNA, 90 days post
         transplant-mouse 3
Lane 7: 950 Rad, spleen DNA, 90 days post
         transplant-mouse 4
Lane 8: 1000 Rad, spleen DNA, 90 days post
         transplant-mouse 5
Lane 9: 1000 Rad, spleen DNA, 90 days post
         transplant-mouse 6
Lane 10: 950 Rad, marrow DNA, 90 day post
         transplant-mouse 3
Lane 11: 950 Rad, marrow DNA, 90 days post
         transplant-mouse 4
Lane 12: 1000 Rad, marrow DNA, 90 days post
         transplant-mouse 5
Lane 13: 1000 Rad, marrow DNA, 90 days post
         transplant-mouse 6

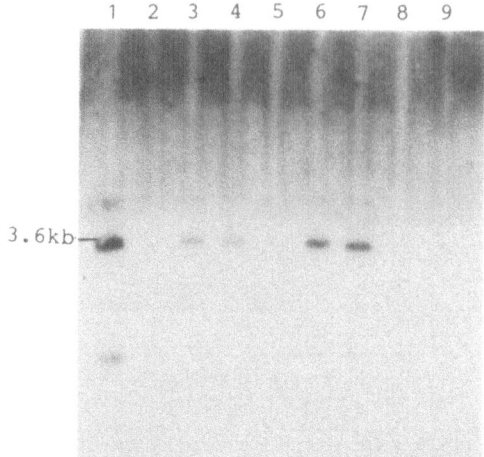

Fig. 3. In order to determine the degree of chimerism bollowing
         transplantation, 10micro gm of DNA are restricted with
         Pvu II and probed using the Y-specific probe (pY2 plasmid)
         obtained from Ed Palmer[31]. Southern blot analysis reveals
         the presence of a 3.6kb band in male mice which is absent
         in female C57Bl/J6 mice. (In the experiment illustrated
         here male mice received female donor marrow.)

         Lane 1:   Y Probe cut with Hinc II
         Lane 2:   Control-Female mouse spleen DNA
         Lane 3:   Control - Male untransplanted mouse spleen
                   DNA
         Lanes 4 and 5: Recipient male spleen DNA from mice
                         transplanted with female marrow with-
                         out prior co-culture with GP+E-N2.
         Lanes 6 and 7: Recipient male spleen DNA from mice
                         transplanted with female marrow in-
                         cubated with GP+E-N2. No N2 demon-
                         strated on Southern Blot analysis
                         following ECO RI digestion.
         Lane 8:   Spleen DNA-male recipient mouse
                   3(N2 shown in Figure 2.)
         Lane 9:   Spleen DNA-male recipient mouse
                   5(N2 shown in Figure 2.)

Chimerism representing repopulation by uninfected female
stem cells following co-culture with the retroviral packaging line, is indicated by the absence of the 3.6kb fragment
in recipient male mice. These mice were negative for the
neomycin resistance gene following digestion with ECO RI
and Southern Blot analysis using the Neo probe. This indicates succesful transplantation of a male mouse with
female stem cells which do not carry the neomycin resistance
gene. On the other hand, chimerism representing a degree
of endogenous male recipient mouse stem cell regeneration,
is indicated by the presence of the 3.6kb band as well
as the 1.5kb Neo fragment thus suggesting the successful
growth of some donor female stem cells carrying the
neomycin resistance gene.

Table 2. Results of Bone Marrow Transplantation With Neomycin Resistance Gene-Titer 1*10⁴

		NUMBER OF MICE SACRIFICED BY DAY SACRIFICED						
RADS	NUMBER	DAY 12	DAY 24	DAY 42	DAY 100	RAD DEATH	LATE DEATH	ALIVE on day 140
1100	18	6	1	2	3	1	3	2
Mice tested		4	1	1	3			
Percent of mice with gene		75%	NK	NK	NK			

NK - Results not known as yet.

Data from southern blot analysis, using Neo probe, of spleen DNA from 9 mice following transplantation with donor marrow incubated with GP+E86 packaging N2 at a titer of 10 particles/ml. IL-1 (obtained from Roche Laboratories) as well as WEHI conditioned medium were used during the 72 hour incubation period.

obtained may reflect the larger piece of DNA inserted or the presence
of inhibitory sequences within the construct. Thus, a second β-Globin
construct, 3.1kb in length, cloned into N2 and extending only 1000bp
3' to the β-Globin gene, is being created. We will soon see whether
this new construct is packaged at higher titers by the GP+E86 packaging
line and is efficiently introduced into mice with acceptable levels of
human β-Globin expression.

In conclusion, our studies to date demonstrate that we have a safe
retroviral packaging line with which we can efficiently transfer simple
small genes into mice. It remains to be seen whether similar results
can be achieved using larger fragments of DNA, including the Human β-Globin
gene, and whether reasonable expression of this gene can be obtained
in the mouse model.

ACKNOWLEDGEMENT

We thank Dr. Eli Gilboa for providing the N2 plasmid and Dr. Dusty
Miller for his gift of the PLNL plasmid.

REFERENCES

1. Cone, R.D., A. Weber-Benarous, D. Baorto, and R.C. Mulligan.
   1987. Regulated expression of a complete human β-Globin gene
   encoded by a transmissable retrovirus vector. Molecular and
   Cell. Biology 7: 887-897.
2. Mann, R., R.C. Mulligan, and D. Baltimore. 1983. Construction
   of a retrovirus packaging mutant and its use to produce helper--
   free defective retrovirus. Cell 33: 153-159.
3. Dick, J.E., M.C. Magli, D. Huszar, R.A. Phillips, and A. Bernstein.
   1985. Introduction of a selectable gene into primitive stem
   cells capable of long-term reconstitution of the hemapoietic
   system of W/W mice. Cell 42: 71-79.
4. Cepko, C.L., B.E. Roberts, and R.C. Mulligan. 1984. Construction
   and applications of a highly transmissable murine retrovirus
   shuttle vector. Cell 37: 1053-1062.
5. Lemischka, I.R., D.H. Raulet, and R.C. Mulligan. 1986. Developmental potential and dynamic behavior of hematopoietic stem
   cells. Cell 45: 917-927.
6. Neel, B.G., W.S. Hayward, H.L. Robinson, J. Fang, and S.M. Astrin.
   1981. Avian leukosis virus-induced tumors have common proviral
   integration sites and sythesize discrete new RNAs: Oncogenesis
   by promoter insertion. Cell 23: 323-334.
7. Varmus, H.E., N. Quintrell, and S. Ortiz. 1981. Retroviruses
   as mutagens: Insertion and excision of a resident nontransforming
   provirus alter expression of a resident transforming provirus.
   Cell 25: 23-36.
8. Cone, R.D., and R.C. Mulligan. 1984. High efficiency gene transfer
   into mammalian cells: Generation of helper-free retrovirus
   with broad mammalian host range. Proc. Natl. Acad. Sci. 81:
   6349-6353.
9. Miller, A.D., M.F. Law, and I.M. Verma. 1985. Generation of
   helper-free amphotropic retroviruses that transduce a dominant-
   acting methotrexate-resistant dihydrofolate reductase gene.
   Mol. Cell. Biol. 5: 431-437.
10. Sorge, J., D. Wright, V.D. Erdman, and A. Cutting. 1984. Amphotropic
    retrovirus system for human cell gene transfer. Mol. Cell.
    Biol. 4: 1730-1737.
11. Watanabe, S., and H.M. Temin. 1983. Construction of a helper

cell line from Avian reticuloendotheliosis virus cloning vectors.
Mol. Cell Biol. 3: 2241-2249.

12. Hock, R.A., and A.D. Miller. 1986. Retrovirus-mediated transfer
and expression of drug-resistant genes in human heamtopoietic
progenitor cells. Nature 320: 275-277.

13. Miller, A.D., and C. Buttimore. 1986. Redesign of retrovirus
packaging cell lines to avoid recombination leading to helper
virus production. Mol. Cell. Biol. 6: 2895-2902.

14. Markowitz D., S. Goff, and A. Bank. 1988. A safe packaging
line for gene transfer: Separating viral genes on two different
plasmids. J. Virol. 62: 1120-1125.

15. Miller, A.D., D.R. Trauber, and C. Buttimore. 1986. Factors involved
in production of helper virus-free retrovirus vectors. Somatic
Cell and Molecular Genetics 2: 175-183.

16. Wright, S. A. Rosenthal, R. Flavell, and F. Grosveld. 1984. DNA
sequences required for regulated expression of β-Globin genes
in murine erythroleukemia cells. Cell 38: 265-273.

17. Chao, M.V., P. Mellon, P. Charnay, T. Maniatis, and R. Axel. 1983.
The regulated expression of β-Globin genes introduced into
mouse erythroleukemia cells. Cell 32: 483-493.

18. Rutherford, T.,and A.W. Nienhuis. 1987. Human globin gene promoter
sequences are sufficient for specific expression of a hybrid
gene transfected into tissue culture cells. Molecular and
Cellular biol. 7: 398-402.

19. Karlsson, S., and A.W. Nienhuis. 1985. Developmental regulation
of human globin genes. Ann. Rev. Biochem. 54: 1071-1108.

20. Costantini, F., G. Radice, J. Magram, G. Stamatoyannopoulous, T.
Papayannopoulou, and K. Chada. 1985. Developmental regulation
of human globin genes in transgenic mice. In: Cold Spring
Harbor Symposium on Quantitative Biology. 50: 361-370.

21. Chada, K., J. Magram, K. Raphael, G. Radice, F. Lacy, and F. Costantini.
1985. Specific expression of a foreign β-Globin gene in erythroid
cells of transgenic mice. Nature 314: 377-380.

22. Charnay, P., R. Treisman, P. Mellon, M. Chao, R. Axel, and T. Maniatis.
1984. Differences in Human A- and β-Globin gene expression
in mouse erythroleukemia cells: The role of intragenic sequences.
Cell 36: 251-264.

23. Towne, T.M., J.B. Lingrel, H.Y. Chen, R.I. Brinster, and R.D. Palmiter.
1985. Erythroid-specific expression of human β-Globin genes
in transgenic mice. EMBO Journal 4: 1715-1723.

24. Lerner, N., S. Brigham, S. Goff,and A. Bank. 1987. Human β-Globin
gene expression after gene transfer using retroviral vectors.
DNA 6: 573-582.

25. Dynam, W.S., and R. Tijan. 1985. Control of eukaryotic messenger
RNA synthesis by sequence specific DNA binding proteins.
Nature 316: 774-778.

26. McKnight, S., and R. Tijan. 1986. Transcriptional selectivity
of viral genes in mammalian cells. Cell 46: 795-805.

27. Cohen, R.B., M. Shefery, and C.G. Kim. 1986. Partial purification
of a nuclear factor that binds to the CAAT box of the mouse
alpha 1 globin gene. Mol. and Cell. Biol. 6: 821-832.

28. Trudel, M., and F. Costantino. 1987. A 3' Enhancer contributes
to the stage-specific expression of the human β-Globin gene.
Genes and Develop. 1: 954-960.

29. Grosveld, F., G. Blom van Assendelft, D.R. Greaves, and G. Kollias.
Position-independant, High-level expression of the human β-Globin
gene in transgenic mice. Cell 51: 975-980.

30. Maniatis, T., P. Fritsch, and J. Sambrook. 1982. A laboratory
manual. Cold Spring Harbor, New York: Cold Spring Harbor Laboratory.

31. Lamar, E.E., and E. Palmer. 1984. Y-Encoded, species specific

DNA in mice: Evidence that the Y Chromosome exists in two polymorphic forms in inbred strains. <u>Cell</u> 37: 171-177.

EXPRESSION OF THE GLOBIN GENES AND HEMATOPOIESIS

IN BETA-THALASSEMIC MICE

Raymond A. Popp[1], Susan L. Bolch[2],
Sarah G. Shinpock[1], and Diana M. Popp[1]

[1]Biology Division, Oak Ridge National Laboratory,
and [2]University of Tennessee-Oak Ridge Graduate
School of Biomedical Sciences, Oak Ridge, TN 37831-8077

Mice homozygous for a deletion of the beta-dmajor globin gene exhibit clinical symptoms of human beta-thalassemia[1,2] and are good experimental animals for investigating the regulation of globin gene expression, perturbation of hematopoiesis, and potential methods for treating patients with beta-thalassemia. Homozygous beta-thalassemic mice have a microcytic anemia, their red blood cells display anisocytosis, poikilocytosis and a shortened life span, and iron overloading occurs in several tissues in response to increased erythropoiesis[2,3]. Mice heterozygous for the beta-thalassemia mutation are clinically normal.

Normal mice have two beta-globin genes.[4,5] Approximately 80 percent of the beta globin in adult mice is produced by the 5' beta-major globin gene and 20 percent by the 3' beta-minor globin gene.[6,7] Homozygous beta-thalassemic mice do not produce any beta-dmajor globin because the entire beta-dmajor globin gene was removed by a spontaneous deletion of a 3.7 kb segment of DNA from mouse chromosome 7.[8] The beta-dminor globin gene was left intact and adult beta-thalassemic mice produce only the beta-dminor globin. Analysis of gene products in mice with several combinations of hemoglobin gene mutations show that the amount of beta globin synthesized in homozygous beta-thalassemic mice is less than in normal mice; however, it is more than would be expected if expression of the beta-dminor globin gene remained at the same level in beta-thalassemic mice as in normal mice. While the amount of beta globin synthesized in beta-thalassemic mice is reduced, a normal quantity of alpha globin is synthesized. After all the beta globin is used to form $\alpha_2\beta_2$ tetramers, the excess alpha globin precipitates. The precipitate damages the erythrocyte membrane, and red cell survival is shortened. To maintain hemostasis, erythropoiesis is expanded. Data on the altered expression of the globin genes and on the perturbed hematopoiesis in beta-thalassemic mice are presented in this report.

MATERIALS AND METHODS

Mice

Mice used in this study inherit mutations at the hemoglobin loci which produce murine thalassemias or are mice that carry unique mutations

Molecular Biology of Erythropoiesis
Edited by J. L. Ascensao *et al.*
Plenum Press, New York

which aid in studies analyzing the expression of the globin genes in thalassemic mice. A spontaneous deletion of the beta-dmajor globin gene occurred in a DBA/2J male and the mutation was recovered from (C57BL/6J X DBA/2J)F1 progeny. Mice that are homozygous for this mutation exhibit clinical symptoms of beta-thalassemia.[1] A congenic strain of beta-thalassemic mice, C57BL/6.D2-$\underline{Hbb^{th}}$, has been developed during the last several years so bone marrow and other hematopoietic tissues can be transplanted between mice of the congenic strain and the inbred partner, C57BL/6J.

Hemoglobin haplotypes

The hemoglobin haplotypes of mice are easily distinguished by electrophoresis of mouse hemoglobin on Titan III cellulose acetate plates (Helena Laboratories, Beaumont, TX). C57BL/6 mice have the beta-single haplotype ($\underline{Hbb^{s}}$), DBA/2 mice have the beta-diffuse haplotype ($\underline{Hbb^{d}}$), and beta-thalassemic mice have the thalassemia haplotype ($\underline{Hbb^{th}}$). Alpha-thalassemic mice that inherit a radiation-induced deletion of both adult alpha-globin genes in the alpha-globin complex[9,10] have the $\underline{Hba^{th}}$ haplotype. Mice that are heterozygous for this mutation have clinical symptoms of alpha-thalassemia,[11] and homozygotes die during early stages of embryogenesis.[12] Mice that inherit a radiation-induced duplication of a segment of chromosome 7, which includes the beta-globin gene complex, have the $\underline{Hbb^{dup}}$ haplotype and carry an extra set of fully functional beta-globin genes.[10] Heterozygotes survive, but the do not. Mice that inherit an ethylnitrosourea-induced mutation in the 5' adult beta-globin gene of the $\underline{Hbb^{s}}$ haplotype have the $\underline{Hbb^{s2}}$ haplotype.[7] The betaglobins encoded by the two adult beta-globin genes of the $\underline{Hbb^{s2}}$ haplotype are easily separated by electrophoresis whereas the two adult beta-globin genes of the $\underline{Hbb^{s}}$ haplotype encode identical polypeptides. These hemoglobin haplotypes and the unique features of their gene products are presented in Table 1.

Hematological indices

Manual methods[13] and an Ortho Diagnostics ELT-15 Hematology Analyzer have been used to obtain data on peripheral blood hematology. The ELT-15 was modified in cooperation with Ortho Diagnostics, Inc. to optimize the instrument for the analysis of mouse blood.[14]

Red cell survival and blood volume

The ^{51}Cr-labelling method was used to measure the rate of red blood cell loss;[15] under homeostatic conditions, this is equal to the rate of red blood cell formation. ^{51}Cr-labelled erythrocytes were also used to determine the blood volumes of normal and beta-thalassemic mice from isotope dilution measurements.[16] Erythrocytes of normal mice were used for the blood volume measurements to avoid the problem of the rapid loss of pitted and deformed red cells that are present in blood of beta-thalassemic mice.

Preparation of bone marrow and spleen cell suspensions

Bone marrow was removed from the four long bones of the hind legs by repeated flushing with 1 ml aliquots of Tyrode's solution (Difco Labs, Detroit, MI) to remove all the bone marrow cells. A single cell suspension was prepared by gentle aspiration of bone marrow through a 27-gauge syringe needle. The spleen was teased apart with forceps and a single cell suspension was prepared by gentle aspiration of small pieces of spleen in Tyrode's solution, initially through a 22-gauge and finally through a 27-gauge syringe needle. Care was taken to obtain maximum recovery of bone marrow and spleen cells because the total cellularity was used to determine the total number of hematopoietic progenitor cells in the bone marrow and spleen of normal and beta-thalassemic mice.

Table 1. List of Globin Haplotypes and their Phenotypes

Haplotype	Adult Globin Genes		Globin Phenotypes
Hbb^s	b1^s	b2^s	Both genes encode identical beta-single globins.
Hbb^d	b1^d	b2^d	b1^d encodes 80% of globin, called beta-dmajor; b2^d encodes 20% of globin, called beta-dminor.
Hbbth	--	b2^d	b1^d gene is deleted; beta-dminor is the only globin synthesized. Homozygotes exhibit beta-thalassemia.
Hbb^{s2}	b1^{s2}	b2^s	b1^{s2} encodes an electrophoretically distinct beta-s2major globin; b2^s encodes the beta-single globin.
Hbb^{dup}	b1^s b2^s; b1^s b2^s		Hbb^s beta-globin complex is duplicated; all four genes are active so beta-globin is produced in excess.
Hba$^{b2(th)}$	--	--	Both alpha-globin genes are deleted. Heterozygotes exhibit alpha-thalassemia; homozygotes die.

Mouse hemoglobin nomenclature is more fully described by Lyon, M.F., Barker, J.E., Popp, R.A. Mouse globin gene nomenclature. J. Hered. 79: 93, 1988.

Hematopoietic stem cell assays

The Till and McCulloch spleen colony technique[17] was used to assess the number of CFU-S in the bone marrow and spleen of normal and beta-thalassemic mice. Single cell suspensions were prepared. The cell concentration was adjusted and 5×10^4, 10^5, 5×10^5, or 10^6 nucleated cells in 0.5 ml of Tyrode's were injected into the lateral tail vein of C57BL/6 recipients that had been exposed to 7 Gy of X-rays (300 kVP, 20 mA, hvl 0.5mm Cu) less than 24 hours before receiving the injection of donor cells. Endogenous spleen colonies did not develop in irradiated controls that did not receive injections of donor cells. Nine or ten days after injection, the recipients were killed by cervical disarticulation and the spleens were removed and placed in Bouin's fixative. Macroscopic colonies were counted and averaged as a single data point for each group of six-to-ten mice that received an injection of the same suspension of cells.

The plasma clot system[18] was used to assess the number of CFU-E in bone marrow and spleen of normal and beta-thalassemic mice. Single cell suspensions were prepared in RPMI 1640 plus 2% FBS and cell concentrations were adjusted to 2-to-6 $\times 10^5$ cells/ml in NCTC-109 (Gibco, Grand Island, NY) containing 20% heat-inactivated FBS (Hyclone, Logan, UT), 1% deionized BSA (Miles Labs, Naperville, IL), 2% beef embryo extract (Gibco), 0.02 mg/ml L-aspargine, 10% bovine citrated plasma (Gibco), 100 units penicillin, 100 mcg streptomycin, 10^{-4} M 2-mercaptoethanol (Sigma, St. Louis, MO), 0.25-0.5 units/ml murine erythropoietin (Hyclone),

and 1 unit bovine thrombin (Sigma) to promote clotting. The mixture
was dispensed in 0.1 ml aliquots in 96-well culture plates and incubated
at 37°C and 5% CO_2 in air for two days. Clots were removed and fixed
on microscope slides, stained with benzidine, and the number of benzidine-
positive colonies containing 8 or more cells in three-to-six clots was
determined at 200-to-600X magnification.

A modified methylcellulose culture medium[19] was used to determine
the number of BFU-E. The medium contained 0.8% methylcellulose in alpha
MEM (Gibco), 1% deionized BSA (Sigma), 20% FBS (Gibco), 10^{-4}M 2-mercapto-
ethanol (Sigma), 100 units penicillin, 100 mcg streptomycin, 1.5-to3.0
units/ml erythropoietin (Connaught-Step II, Swiftwater, PA), 40 units
IL-3/ml (a gift of Dr. J. Ihle, Fredrick's Cancer Institute, Fredrick,
MD), and 0.02 mM hemin was added to enhance the number and size of the
developing BFU-E.[20] In some experiments, premixed HCC-10.α3a medium
(Terry Fox Labs, Vancouver, Canada) was used. Cultures were plated with
5×10^4-to-2×10^5 bone marrow cells or with 2-to-6×10^5 spleen cells
in 1 ml volumes in 6-well culture plates (Flow Labs, McLean, VA) and
incubated at 37°C and 5% CO_2 in air for 8 or 9 days. The number of colonies
with more than 50 cells that contained hemoglobin was determined in triplicate
wells at 40-to-100X magnification. Both culture media gave similar results,
so the data were pooled.

A modified agar culture system[21] was used to assay the number of
GM-CFC (CFU-C). The medium contained 0.3% agar (Difco), 15% FBS (Gibco),
5% horse serum (Gibco), and 0.02 mg/ml L-asparagine in supplemented alpha
MEM (Gibco). Two colony stimulating factors (CSF) were used in these
experiments, either appropriately diluted (usually 1:7) 6 hr post-endotoxin
mouse serum (ET)[22] or 10% pokeweed-mitogen spleen-cell-conditioned medium
(PWMCM).[23] Cultures were plated with 7.5×10^4 bone marrow cells or
with 2×10^5 spleen cells in 1 ml volumes in 6-well culture plates (Flow
labs) and incubated at 37°C and 7.5% CO_2 in air for 7 days. Colonies
with more than 50 cells were counted in triplicate wells at 40-to-100X
magnification.

Data from the hematopoietic stem cell assays will be presented both
as the number of progenitor cells per 10^5 bone marrow and spleen cells
and as the total number of progenitor cells per tissue. There is a large
difference between the cellularity of the spleens of normal and beta-thalassemic mice. The absolute numbers of progenitor cells are more important
for interpretation of total hematopoiesis, but the data on the incidence
of progenitor cells indicates the numbers of progenitor cells in the
samples analyzed.

Quantitation of peripheral blood hemoglobins

The relative amounts of allelic forms of hemoglobin in circulating
erythrocytes were determined by electrophoresis on Titan III cellulose
acetate plates[2] and by quantifying the hemoglobin bands with a Flur-Vis
transmission densitometer (Helena Labs Autoscan, Beaumont, TX) equipped
with an electronic integrator. The two kinds of erythrocytes present
in fetuses at 14.5 days of gestation were separated by centrifugal elutriation[24] in order to quantitate the relative amounts of the multiple forms
of adult hemoglobins in erythrocytes that develop in the visceral yolk
sac versus those that develop in the fetal liver.

The quantity of globin produced from a single gene can be determined
by multiplying the percentages of each hemoglobin in the hemolysate times
the MCH. The average MCH value for erythrocytes of normal mice ranges
from 12-to-15 pg/red blood cell; 13 pg/red blood cell will be used for
these calculations.

In vitro synthesis of globins

The relative synthesis of the alpha and beta globins in peripheral blood reticulocytes and bone marrow from normal and thalassemic mice was determined by the incorporation of ^{3}H-leucine and by separation of the globins over columns of carboxymethylcellulose.[25]

Statistical analysis

The student's t test was used to examine the level of significance between data obtained from normal and beta-thalassemic mice. Linear regression analysis was used to compute the $t\frac{1}{2}$ values for the survival of circulating erythrocytes.

RESULTS

Hematological indices

Data on peripheral blood hematology (Table 2 and ref. 2) showed that homozygous beta-thalassemic mice have significantly lower hematocrits (HCT), reduced red cell counts (RBC), lower values for hemoglobin (HGB), mean corpuscular hemoglobin (MCH), mean corpuscular volume (MCV), and mean corpuscular hemoglobin concentration (MCHC), and elevated reticulocyte (RET) and nucleated cell counts (WBC) than normal and heterozygous beta-thalassemic mice. These data established that homozygous beta-thalassemic mice have a hypocellular, hypochromic, and microcytic anemia.

Table 2. Summary of Hematologic Findings

Peripheral Blood:	Normal	Beta-Thal
WBC (x 10^3/ml)	6.1	16.8
RBC (x 10^9/ml)	12.1	9.9
RET (%)	2.0	20.0
HGB (g/dl)	15.3	9.0
HCT (%)	47.5	32.8
MCV (fl)	41.8	35.6
MCH (pg/rbc)	12.7	9.1
MCHC (g/dl)	32.2	27.4
Blood volume (ml)	1.45	1.50
RBC half life (days)	15.0	8.0
Daily RBC production (x 10^8)	8.1	12.9[a] 15.9[b]

[a]Calculation based on the adjusted linear regression t1/2 of 8 days. See footnote to Figure 2. Daily RBC production = daily RBC loss = RBC (cells/ml) x blood volume (ml) x $\dfrac{1}{\text{RBC mean life span}}$.

RBC mean life span = $\dfrac{t\ 1/2}{0.693}$.

[b]Calculation based on the non-adjusted linear regression t1/2 of 6.5 days. See footnote to Figure 2.

The high reticulocyte but low red cell counts in beta-thalassemic mice suggested that the production of erythrocytes was higher than normal but the erythrocytes were also being removed from the peripheral circulation more rapidly than in normal mice. Scanning electron micrographs showed that an occasional red cell was misshapen or pitted in blood of heterozygous beta-thalassemic mice, but in the blood of homozygous beta-thalassemic mice the majority of the red cells were small, irregular in shape, or contained less than normal amounts of hemoglobin (Figure 1).

Red cell survival

Analyses on the survival of ^{51}Cr-labelled erythrocytes in normal and beta-thalassemic mice (Figure 2) revealed that the half life of beta-thalassemic red cells was significantly less than normal red cells. Erythrocytes from normal mice had a half life of 15 days when transfused into normal recipients and a half life of 23.5 days in homozygous beta-thalassemic recipients. Erythrocytes from heterozygous beta-thalassemic mice had a half life of 14 days in normal recipients. Erythrocytes from homozygous beta-thalassemic mice had a longer half life in splenectomized (11 days) than in intact beta-thalassemic (8 days) and normal (6.75 days) recipients. These results established that erythrocytes from beta-thalassemic mice have a shortened life span irrespective of the host in which the red cells were circulating.

Under homeostatic conditions, the number of red cells lost each day are replaced by newly formed red cells. The blood volumes of adult mice were determined in order to calculate the total production of erythrocytes in normal and beta-thalassemic mice. Normal and beta-thalassemic mice at 3-to-4 months of age and weighing 25g had blood volumes of 1.45 ± 0.04 and 1.50 ± 0.04 ml, respectively, and the average red cell counts in these mice were 12.1 and 9.9 x 10^9/ml, respectively. The daily red cell production was calculated[15] to be 8.1 x 10^8 for normal mice and 12.9-to-15.9 x 10^8 for beta-thalassemic mice. These data established that beta-thalassemic mice produce more red cells each day than normal mice.

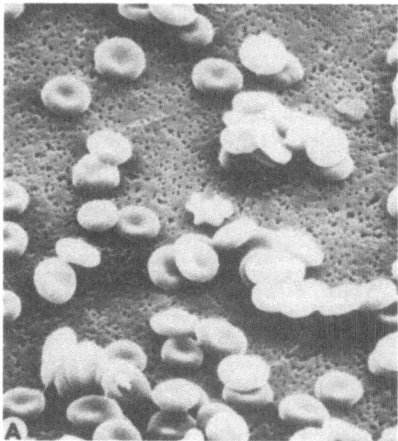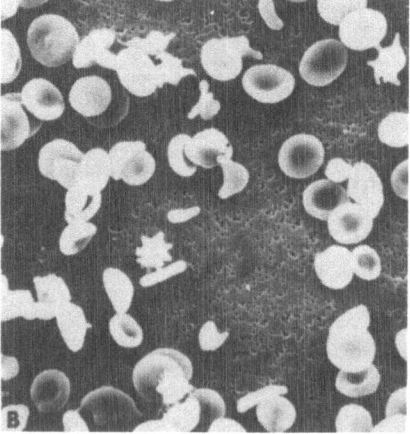

Fig. 1.   Scanning electron micrographs of red blood cells from beta-thalassemic mice. A. Erythrocytes from a heterozygous beta-thallassemic mouse. Magnification 2000X. B. Erythrocytes from a homozygous beta-thalassemic mouse. Magnification 2000X.

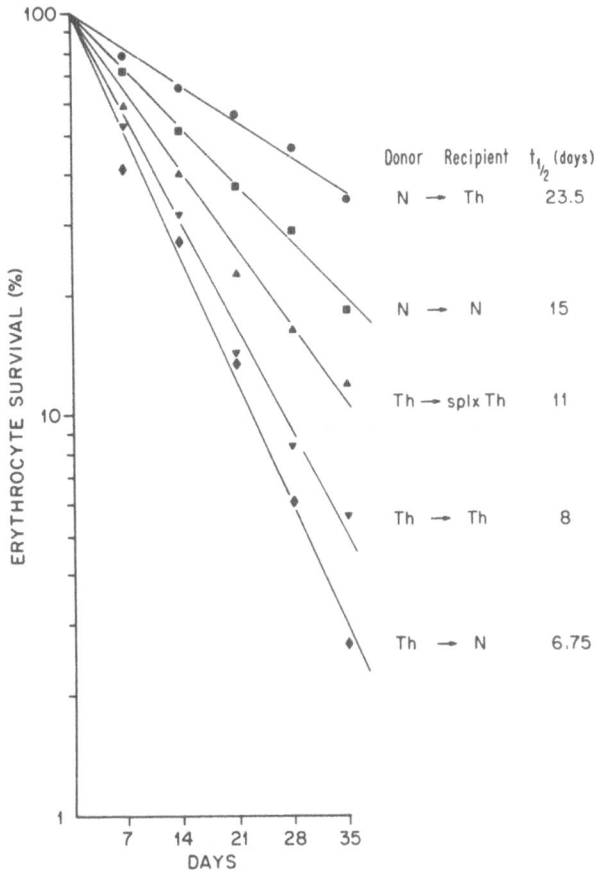

Fig. 2. Erythrocyte survival in normal and beta-
thalassemic mice. Ten microliter samples
of blood were removed at each bleeding. All
counts were expressed as a percentage of the
count in ten microliters of blood removed at
24 hours after the ^{51}Cr-labelled erythrocytes
were injected. The least squares fit of the
Th → splxTh and Th → lines intercepted the
day 0 Y-axis at 85 rather than 100 percent.
The difference was due to the more rapid
destruction of the most defective erythrocytes. These lines were adjusted upward to
intercept the Y-axis at 100 percent. The
non-adjusted t1/2 values were 8.5 days for
the → splxTH and 6.5 days for Th → Th.

Hematopoietic stem cell assays

In the bone marrow of beta-thalassemic mice, the incidence of CFU-E
was about two-fold higher than in normal mice but the incidences of BFU-E,
CFU-C and CFU-S were similar for normal and beta-thalassemic mice (Figure
3). The total cellularity of the bone marrow from beta-thalassemic mice
was only slightly higher than from normal mice (Table 3). Thus, the
total number of CFU-E was 2.2-fold higher in beta-thalassemic mice, but
the total numbers of BFU-E, CFU-C and CFU-S in the bone marrow of the
two groups of mice was similar (Table 3).

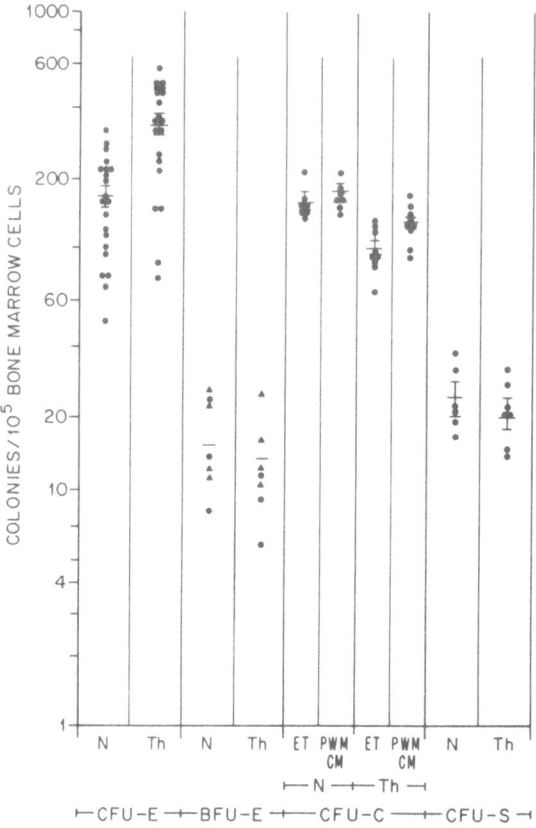

Fig. 3.   Hematopoietic progenitor cells in the bone
marrow. The number of colonies that developed
from 10^5 bone marrow cells of normal (N) and
beta-thalassemic (Th) mice are compared. ET,
endotoxin-stimulation growth factor; PWMCM,
pokeweed-mitogen spleen-cell-conditioned
medium.

In spleens of beta-thalassemic mice, the incidences of CFU-E and
CFU-C were significantly higher than in the spleens of normal mice, but
the incidences of BFU-E and CFU-S were similar in the spleens of the
two groups of mice (Figure 4). Both spleen weight and cellularity were
significantly higher in beta-thalassemic than in normal mice. Thus,
there was a significant increase in the total numbers of the hematopoietic
progenitor cells tested in the spleens of beta-thalassemic mice (Table
3). These data established that the spleen provides most of the expanded
erythropoiesis required to sustain erythroid homeostasis in beta-thalassemic
mice.

Peripheral blood hemoglobins in beta-thalassemic mice

The proportions and quantities of the variant forms of hemoglobin
tetramers in hemolysates of peripheral blood of normal and beta-thalassemic
mice are presented in Table 4. Compared with the 1.3 pg of beta-dminor
hemoglobin in erythrocytes of normal Hbb^s/Hbb^d heterozygotes, the quantity
of the beta-dminor gene product synthesized in each red cell was increased
to 3.2 pg in erythrocytes of beta-thalassemic Hbb^s/Hbb^th heterozygotes.
Compared with the 2.6 pg of beta-dminor hemoglobin in erythrocytes of

Table 3. Total Hematopoietic Progenitor Cells in Bone Marrow and Spleen
of Normal and Beta-Thalassemic Mice

Tissue[a] (No. animals)	Cellularity[b] ($\times 10^6$) $\pm$ SEM	Progenitor cells ($\times 10^3$)			
		CFU-E	BFU-E	CFU-C	CFU-S
NBM (11)	63.3 $\pm$2.0	113.8	10.0	92.9	15.3
ThBM (13)	71.3[c] $\pm$2.2	245.1	10.0	84.8	14.3
N Spleen (16)	176.3 $\pm$10.1	15.9	18.1	4.2	2.8
Th Spleen (17)	721.2[d] $\pm$38.3	4803.2	27.4	151.5	17.3

[a] NBM, normal mouse bone marrow; ThBM, thalassemic mouse bone marrow; N
Spleen, normal mouse spleen; Th Spleen, thalassemic mouse spleen.

[b] Bone marrow cells in two femors and two tibias. Spleen cells in one
spleen.

[c] $p < 0.025$

[d] $p < 0.0001$

normal Hbb^d/Hbb^d mice, the quantity of the beta-dminor hemoglobin was
increased to 5.9 and 9.0 pg in erythrocytes of heterozygous beta-thalassemic
(Hbb^d/Hbbth) and homozygous beta-thalassemic (Hbbth/Hbbth) mice, respectively. The data from heterozygotes suggested that up-regulation of the
beta-dminor globin gene was controlled by cis-acting factors because
the quantity of globin encoded by the Hbbth allele was increased by 200-to-
-300 percent, whereas the products of the Hbb^s, Hbb^d and Hbb^{s2} alleles
were increased by only 25-to-35 percent. In mice of the Hbb^{s2}/Hbbth
genotype, the quantity of the beta-sminor globin was increased from 2.34
to 2.99 pg/rbc, a 28 percent increase, which is equivalent to the 35
percent increase of beta-s2major globin from 4.42 to 5.98 pg/rbc. In
these mice, the quantity of the beta-dminor globin was increased from
1.43 to 4.03 pg/rbc, a 282 percent increase.

The time during development when the beta-dminor globin gene became
up-regulated in beta-thalassemic mice was determined by analyzing the
relative quantities of the beta-s2major, beta-sminor and beta-dminor
globins in erythrocytes of Hbb^{s2}/Hbbth fetuses at 13.5-to-18.5 days of
gestation (Table 5). Data on the percentage of beta-dminor hemoglobin
in separate hemolysates of nucleated erythrocytes that differentiate
in the visceral yolk sac blood islands and of non-nucleated erythrocytes
that develop in the fetal liver established that the beta-dminor globin
gene was already up-regulated by 14.5 days of gestation.

Mice of the Hbb^{s}Hbb^s/Hbb^d genotype were used to determine what effect
an extra set of beta-globin genes would have on the relative quantities
of the various hemoglobins synthesized and on the expression of the duplicated set of beta-single globin genes. Twice as much beta-single hemoglobin
as beta-dmajor plus beta-dminor hemoglobins was found in peripheral blood.
These data indicated that the two beta-globin genes in both of the Hbb^s

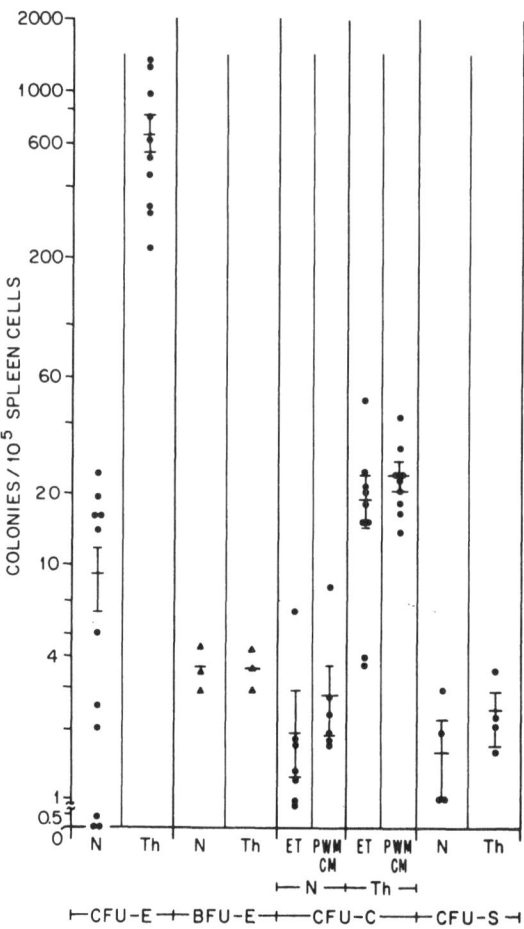

Fig. 4. Hematopoietic progenitor cells in the spleen.
For explanation see legend for Figure 3.

complexes were equally as active as the two beta-globin genes in the
Hbb^d complex on the opposite chromosome (Table 4).

In vitro synthesis of globins in thalassemic mice

Data on the ratios of alpha and beta globins synthesized in peripheral
blood reticulocytes and bone marrow are presented in Table 6. Beta-globin
synthesis was somewhat depressed (β/α = 0.92) in peripheral blood reticulocytes from heterozygous beta-thalassemic mice. However, beta-globin
synthesis was markedly less than alpha-globin synthesis in peripheral
blood reticulocytes (β/α = 0.75) and bone marrow (β/α = 0.58) of homozygous
beta-thalassemic mice. Mice that carry a duplication of the beta-single
globin complex, Hbb^{S}Hbb^S/Hbb^d, synthesized only slightly more beta globin
than alpha globin (β/α = 1.09) but alpha-thalassemic mice that carry
a deletion of one alpha-globin complex exhibited a significant deficiency
of alpha-globin synthesis (β/α = 1.35). The synthesis of alpha and beta
globins was nearly equivalent in doubly heterozygous alpha-, beta-thalassemia
mice (β/α = 0.96) and in mice that were homozygous for beta-thalassemia
as well as heterozygous for alpha-thalassemia (β/α = 0.93). These data
established that the relative amounts of beta globin synthesized in beta-
thalassemic and beta-duplication mice, and of alpha globin synthesized

Table 4. Quantity of Various Kinds of Hemoglobin in Blood of Mice of Different Genotypes

Hemoglobin genotype	Percent				MCH (pg/rbc)[a]			
	Beta-single major	minor	Beta-diffuse major	minor	Beta-single major	minor	Beta-diffuse major	minor
Hbb^S/Hbb^S	100				13.0			
Hbb^S/Hbb^d	50		40	10	6.5		5.2	1.3
Hbb^d/Hbb^d			80	20			10.4	2.6
Hbb^S/Hbbth	75			25	9.8			3.2
Hbb^d/Hbbth			55	45			7.2	5.9
Hbbth/Hbbth			100					9.0
Hbb^{s2}/Hbb^{s2}	71	29			9.2	3.8		
Hbb^{s2}/Hbb^d	34	18	37	11	4.4	2.3	4.8	1.4
Hbb^{s2}/Hbbth	46	23		31	6.0	3.0		4.0
Hbb^{S}Hbb^S/Hbb^d		67	26	7	8.7		3.4	0.9

[a]The calculated MCH for the normal and heterozygous beta-thalassemic genotypes in this table ranged from 12.1 to 13.6 pg/rbc and was 9.0 pg/rbc for homozygous beta-thalassemic mice. In order to compare these groups, the hemoglobin in erythrocytes of normal and heterozygous beta-thalassemic mice was normalized to 13 pg/rbc.

in alpha-thalassemic mice, were closer to normal than might have been expected based on the dosage of active beta- and alpha-globin genes in mice that carry these mutations.

DISCUSSION

These studies confirmed that many hematological and clinical features of murine and human beta-thalassemia are similar. Mice that are heterozygous carriers were asymptomatic, but mice homozygous for the deletion of beta--dmajor globin gene exhibited a hypocellular, hypochromic and microcytic anemia (Table 2). The anemia was caused by a combination of several factors that originate from the mutation. Deletion of the beta-dmajor globin gene resulted in the synthesis of less beta globin while synthesis of alpha globin proceeded normally (Table 6). The reduced quantity of hemoglobin was packaged in smaller erythrocytes. After all the beta globin was assembled into $\alpha_2\beta_2$ tetramers, the alpha globin that remained precipitated. The precipitate damaged the red cell membrane (Figure 1). These defective erythrocytes were removed more rapidly from the peripheral circulation so the mean survival of erythrocytes from beta-thalassemic mice exhibited a shorter than normal life span (Figure 2). Splenectomy extended the life span of the defective erythrocytes in the circulation of beta-thalassemic mice.

The difference between the β/α-globin synthesis in peripheral blood reticulocytes and in bone marrow (Table 6) suggests that the imbalance

Table 5. Quantity of Various Kinds of Hemoglobin in Fetal Blood of Heterozygous Beta-Thalassemic Mice

Day of gestation	Percent		
	Beta-single		Beta-diffuse
	major	minor	minor
13.5	32.1[a]	38.4	29.5
14.5 (nucleated)	25.0[a]	43.9	31.0
14.5 (non-nucleated)	36.6	31.8	31.6
15.5	66.6[b]		33.4
16.5	65.3[b]		34.7
18.5	42.7[a]	26.7	30.7
Adult	46.0[a]	23.0	31.0

[a] Hbb^{s2}/Hbb^{th} heterozygotes; the products of the beta-single major and minor globin genes can be distinguished.

[b] Hbb^{s}/Hbb^{th} heterozygotes; the products of the beta-single major and minor globin genes cannot be distinguished.

Blood from each fetus was collected and analyzed separately except for the 14.5-day fetuses where the blood from 8 fetuses was pooled and the nucleated, visceral yolk sac and non-nucleated, fetal liver erythrocytes were separated by centrifugal elutriation prior to analysis of their hemoglobins.

of alpha- and beta-globin synthesis might be greater in reticulocytes in the bone marrow than in peripheral blood. The data are consistent with the possibility that some of the reticulocytes in the bone marrow were destroyed before they entered the peripheral circulation. Similar dysfunctional erythropoiesis has been described for beta-thalassemia patients.[26]

This mouse model for human beta-thalassemia permitted us to investigate how the numbers of hematopoietic stem cells and/or erythroid progenitor cells are increased to compensate for the shortened life span of the defective erythrocytes and the dysfunctional erythropoiesis in beta-thalassemic mice. The results (Table 3 and Figure 3) showed that cellularity of the bone marrow was increased by 12 percent and the total number of CFU-E progenitor cells was 215 percent greater than for normal bone marrow. In the spleen, there was a four-fold increase in cellularity as well as an increase in the incidence of CFU-E and CFU-C progenitor cells (Table 3 and Figure 4). The largest expansion occurred in the CFU-E compartment; the incidence of CFU-E was increased 74-fold and the total number was increased 300-fold (Table 3). In contrast, the BFU-E compartment in the spleen expanded less than the increase in cellularity.

Our data showed that there was a selective expansion of the CFU-E compartment in the spleen of beta-thalassemic mice. A similar increase in the incidence of CFU-E but not BFU-E in mice has been reported in experiments in which the serum erythropoietin levels were elevated by bleeding or by the injection of erythropoietin.[27] Although erythropoietin

Table 6. Relative Synthesis of Alpha and Beta Globins in Thalassemic Mice

Hemoglobin types	No. of analysis		Beta/alpha globin synthesis	No. of beta/alpha globin genes
Normal	15	PB[a]	1.00[c]	4/4
Heterozygous beta-thal	3	PB	0.92±0.04	3/4
Homozygous beta-thal	9	PB	0.75±0.04	2/4
Homozygous beta-thal	3	BM[b]	0.58±0.09	2/4
Beta-duplication	1	PB	1.09	6/4
Heterozygous alpha-thal	6	PB	1.35±0.06	4/2
Doubly heterozygous alpha-, beta-thal	3	PB	0.96±0.01	3/2
Heterozygous alpha-, homozygous beta-thal	3	PB	0.93±0.01	2/2

[a]Peripheral blood reticulocytes

[b]Bone marrow

[c]The ratio of beta/alpha-globin synthesis was 0.99 ± 0.02 for the 15 control samples. This average was assigned the value of 1.00. A separate control was done for each experiment and its beta/alpha-globin synthesis was assigned the value of 1.00; the beta/alpha-globin synthesis from mice of other genotypes was compared to its own control (normalized) to allow comparison of the results of experiments done at different times.

levels in beta-thalassemic mice have not been determined, an increased quantity of erythropoietin would be consistent with the marked increase in the incidence of CFU-E and not BFU-E.

One would not expect that expansion of the CFU-C progenitor cells would be associated with an expansion of the CFU-E compartment. However, the incidence of splenic CFU-C was 8.75-fold above normal in beta-thalassemic mice (Figure 4) where the incidence of CFU-E was increased 74-fold. The total number of splenic CFU-C progenitor cells was increased 36-fold and the total number of CFU-E was increased 300-fold above normal in beta-thalassemic mice (Table 3). These cells are decendants of a common myeloid stem cell but their growth is governed by different growth factors. Moreover, erythropoietin has not been shown to induce the growth of CFU-C nor to modify CFU-C morphology.[28] The splenic CFU-E and CFU-C compartments were increased more in beta-thalassemic mice (Table 3) than in alpha-thalassemic mice.[29,30]

Increased erythropoiesis alone would not be sufficient to provide the quantity of hemoglobin necessary for these mice to survive. Our studies have established that the intact beta-dminor globin gene in beta-thalassemic mice is up-regulated 200-to-300 percent in a cis-activating manner (Tables 4 and 5). The beta-dmajor globin gene naturally lies about 14 kb 5' to the beta-dminor globin gene[31] and the 3.7 kb deletion breakpoints were located at nucleotide -1984 5' to the cap site and between nucleoties +1727 and 1731 3' to the cap site.[8] Thus, nearly all of the 14 kb segment between the beta-dmajor and beta-dminor genes has remained intact and the <u>cis</u>-acting sequences responsible for this up-regulation may be located in this 14 kb segment or 5' to the deletion. Although <u>cis</u>-acting promotor and enhancer sequences are usually thought to lie near the coding sequences, it has recently been shown that constructs of the human alpha- and beta-globin genes which included DNAse hypersensitivity sites I and II located 5' to the epsilon-globin gene more than 50 kb 5' to the beta-globin gene were capable of being expressed at high efficiency in transgenic mice.[32,33] In contrast, the alpha- and beta-globin genes alone were always poorly expressed. Sequences responsible for the up-regulation of the beta-dminor globin gene in beta-thalassemic mice have not been identified. In concert with increased hematopoiesis, up-regulation of the beta-dminor globin gene enables the beta-thalassemic mouse to produce a sufficient amount of hemoglobin to survive.

We have established that the beta-thalassemic mouse is an excellent animal model for human beta-thalassemia. Two congenic lines have been established; one on the C57BL/6J genetic background that was used for these studies and the other on the DBA/2J genetic background. We believe that these mice are excellent candidates for cell replacement and gene therapy. The nucleotide sequences adjacent to the deletion breakpoints have been determined[8] so site-directed gene insertion procedures described today[34] might be tested using bone marrow cells from these beta-thalassemic mice.

Viral vectors have proved to be the most efficient method to transfer exogenous genes into the DNA of bone marrow stem cells,[35,36,37,38] despite the fact that the integrated globin genes are poorly expressed. Viral vectors that include the human beta-globin gene and the 5' DNAse hypersensitivity sites are being constructed by others[38] and their efficacy for gene therapy will be tested using bone marrow cells from beta-thalassemic mice.

We have compared the values for various progenitor cell compartments in normal and beta-thalassemic mice. It is assumed that the perturbed hematopoiesis in beta-thalassemic mice would return to near normalcy if beta-globin synthesis could be increased significantly by gene therapy. The base values given in this paper for normal and beta-thalassemic mice can be used as a reference to judge the efficacy of foreign globin genes to return hematopoiesis in beta-thalassemic mice to normalcy.

ACKNOWLEDGEMENTS

The authors thank Drs. C.J. Wawrzyniak and J.H. Dumont for making the scanning electron micrographs, N. Crowe for preparing the illustrations, and P. Lyons and A. Barnes for typing the manuscript.

Research sponsored jointly by NIH Grant 5 RO1 HL 37056, NIH Predoctoral Training Grant in Genetics GM-7438 to SLB, and the Office of Health and Environmental Research, U.S. Department of Energy under Contract DE-AC05-84OR21400 with Martin Marietta Energy Systems, Inc.

REFERENCES

1. Skow, L.C., B.A. Burkhart, F.M. Johnson, R.A. Popp, D.M. Popp, S.Z. Godberg, W.F. Anderson, L.B. Barnett, and S.E. Lewis. 1983. A mouse model for β-thalassemia. <u>Cell</u> 34: 1043.
2. Popp, R.A., D.M. Popp, F.M. Johnson, L.C. Skow, and S.E. Lewis. 1985. Hematology of a murine β-thalassemia: A longitudinal study. <u>Ann. NY Acad. Sci.</u> 445: 432.

3. Van Wyck, D.B., M.E. Tancer, and R.A. Popp. 1987. Iron homeostasis in β-thalassemic mice. Blood 70: 1462.

4. Tiemeier, D.C., S.M. Tilghman, F.I. Polsky, J.G. Seidman, A. Leder, M.H. Edgell, and P. Leder. 1978. A comparison of two cloned mouse β-globin genes and their surrounding and intervening sequences. Cell 14: 237.

5. Weaver, S., M.B. Comer, C.L. Jahn, C.A. Hutchison, III, and M.H. Edgell. 1981. The adult β-globin genes of the "single" type mouse C57BL. Cell 24: 403.

6. Popp, R.A., and E.G. Bailiff. 1973. Sequence of amino acids in the major and minor β-chains of the diffuse hemoglobin from BALB/c mice. Biochim. Biophys. Acta. 303: 61.

7. Wawrzyniak, C.J., and R.A. Popp. 1985. Use of a new mouse beta-globin haplotype (Hbb^{s2}) to study hemoglobin expression during development. Dev. Biol. 112: 477.

8. Goldberg, S.Z., D. Kuebbing, D. Trauber, M.P. Schafer, S.E. Lewis, R.A. Popp, and W.F. Anderson. 1986. A 66-base pair insert bridges the deletion responsible for a mouse model for βthalassemia. J. Biol. Chem. 261: 12368.

9. Russell, L.B., W.L. Russell, R.A. Popp, C. Vaughan, and K.B. Jacobson. 1976. Radiation-induced mutations at the mouse hemoglobin loci. Proc. Natl. Acad. Sci. USA. 73: 2843.

10. Popp, R.A., L.P. Stratton, D.K. Hawley, and K. Effron. 1979. Hemoglobin of mice with radiation-induced mutations at the hemoglobin loci. J. Mol. Biol. 127: 141.

11. Popp, R.A., and M.K. Enlow. 1977. Radiation-induced α-thalassemia in mice. Am. J. Vet. Res. 38: 569.

12. Popp, R.A., B.S. Bradshaw, and L.C. Skow. 1980. Effects of alpha thalassemia on mouse development. Differentiation 17: 205.

13. Wintrobe, M.M., G.R. Lee, D.R. Boggs, T.C. Bithell, J. Forester, J.W. Athens, and J.N. Lukens (eds). 1981. in: "Clinical Hematology." 8th Edition. Lea and Febiger, Philadelphia, PA.

14. Popp, D.M., R.A. Popp, S. Lock, R.C. Mann, and J.R. Hand. 1986. The use of multiparameter analysis to quantitate hematological damage from exposure to a chemical (ethylene oxide). J. Toxicol. Environ. Health. 18:543.

15. The International Committee for Standarization in Hematology. 1971. Recommended methods for radioisotope and red cell survival studies. Blood. 38: 378.

16. Paxson, C.L., and L.H. Smith. 1968. Blood volume of the mouse. Exp. Hematol. 17: 42.

17. Till, J.E., and E.A. McCulloch. 1961. A direct measurement of the radiation sensitivity of normal mouse bone marrow. Radiat. Res. 14: 213.

18. McLeod, D.L., M.M. Shreeve, and A. Axelrad. 1974. Improved plasma clot system for production of erythrocytic colonies in vitro: Quantitative assay method for CFU-E. Blood 44: 517.

19. Iscove, N.N., and F. Sieber. 1975. Erythroid progenitors in mouse bone marrow detected by macroscopic colony formation in culture. Exp. Hematol. 3: 32.

20. Porter, P.N., R.H. Meints, and K. Mesner. 1979. Enhancement of erythroid colony growth in culture by hemin. Exp. Hematol. 7: 11.

21. Metcalf, D. 1977. Hemopoietic Colonies. Springer-Verlag, New York, NY.

22. Metcalf, D. 1971. Acute antigen-induced elevation of serum colony stimulating factors (CFS) levels. Immunology 21: 427.

23. Parker, J.W., and D. Metcalf. 1974. Production of colony-stimulating factor in mitogen-stimulated lymphocyte cultures. J. Immunol. 112: 502.

24. Wawrzyniak, C.J., and R.A. Popp. 1987. Expression of the two adult beta-globin genes in mouse yolk sac and fetal liver erythrocytes. Dev. Biol. 119: 299.

25. Martinell, J., J.B. Whitney III, R.A. Popp, L.B. Russell, and W.F. Anderson. 1981. Three mouse models of human thalassemia. Proc. Natl. Acad. Sci. USA. 78: 5056.

26. Weatherall, D.J., and J.B. Clegg. 1981. The Thalassemia Syndromes, 3rd ed. Blackwell Scientific Publications, Oxford, England.

27. Hara, H., and M. Ogawa. 1977. Erythropoietic precursors in mice under erythropoietic stimulation and suppression. Exp. Hematol. 5: 141.

28. Metcalf, D. 1969. The effect of bleeding on the number of in vitro colony-forming cells in the bone marrow. Brit. J. Haemat. 16: 397.

29. Barker, J.E., and E. McFarland. 1985. The hematopoietic stem cells of α-thalassemic mice. Blood 66: 595.

30. Wagemaker, G., and T.P. Visser. 1986. Enumeration of stem cells and progenitor cells in alpha-thalassemic mice reveals lack of specific regulation of stem cell differentiation. Exp. Hematol. 14: 303.

31. Jahn, C.L., C.A. Hutchison III, S.J. Phillips, S. Weaver, N.L. Haigwood, C.F. Voliva, and M.H. Edgell. 1980. DNA sequence organization of the β-globin complex in the BALB/c mouse. Cell 21: 159.

32. Grosveld, F., G.B. van Assendelft, D.R. Greaves, and G. Kollias. 1987. Position-independent, high-level expression of the human β-globin gene in transgenic mice. Cell 51: 975.

33. Townes, T. 1988. High level erythroid specific expression of human globin genes in transgenic mice: Presented at the Sixth Conference on Hemoglobin Switching. Held at Airlie, VA, September 24-27, 1988.

34. Popovich, B. 1988. Targeted modification of the beta-globin gene. Presented at the Fourth Symposium on Molecular Biology of Erythropoiesis. Held at Reno, NV, October 30-November 2, 1988.

35. Williams, D.A., I.R. Lemischka, D.G. Nathan, and R.C. Mulligan. 1984. Introduction of a new genetic material into pluripotent haematopoietic stem cells of the mouse. Nature 310: 476.

36. Li, C.L. 1988. Human beta-globin viral vectors. Presented at the Fourth Symposium on Molecular Biology of Erythropoiesis. Held at Reno NV, October 30-November 2, 1988.

37. Hesdorffer, C. 1988. Gene transfer into bone marrow cells. Presented at the Fourth Symposium on Molecular Biology of Erythropoiesis. Held at Reno, NV, October 30-November 2, 1988.

38. Gelinas, R. 1988. Regulated expression of human beta-globin RNA and protein in hematopoietically reconstructed mice. In: Fourth Symposium on Molecular Biology of Erythropoiesis. Held at Reno NV, October 30-November 2, 1988.

BUTYRIC ACID MODULATES DEVELOPMENTAL GLOBIN GENE

SWITCHING IN MAN AND SHEEP

Susan P. Perrine, Paul Swerdlow, Douglas V. Faller, Gene Qin,
Abraham M. Rudolph, James Reczek, and Yuet Wai Kan

Children's Hospital Oakland Research Institute, Oakland, CA;
Medical College of Virginia, Richmond, VA; Dana Farber Cancer
Institute, Boston, MA; Howard Hughes Medical Institute, Depart-
ments of Medicine and Pediatrics, University of California,
San Francisco, CA.

ACKNOWLEDGEMENTS

Supported by NIH grants HL-37118, HL-20985, and by a grant-in-aid
from the American Heart Association, California Affiliate, with funds
contributed by the Alameda County Chapter.

ABSTRACT

The developmental switch from production of fetal (γ) to adult (β)
globin occurs on a normally set biologic clock which proceeds even if
the adult (β) globin genes are defective. Preventing or reversing the
globin gene switch would be beneficial for subjects with abnormal β globin
genes. We have now identified a class of agents which, when present
in elevated plasma concentrations during gestation, appears to inhibit
the $\gamma \rightarrow \beta$ globin gene switch in developing humans. Further investigation
has shown that butyric acid and related compounds can increase γ globin
and decrease β globin expression in erythroid cells cultured from subjects
with diseases of abnormal β globin. Butyrate compounds were therefore
infused in an in vivo fetal animal model, and the globin switch was inhibited
in most and reversed in some fetal lambs. These data suggest that inhibiting
expression of abnormal β globin genes may be possible in future generations.
Histone modification may be a mechanism of action involved.

The developmental switch from production of γ globin to β globin
results in significant morbidity when the β globin genes are defective.
The globin switch has therefore been extensively studied, appearing to
be set on a biologic clock and proceeding despite the site of blood production and solely on the basis of gestational age[1]. We previously found
that this developmental gene switch is delayed in human fetuses developing
in the presence of maternal diabetes[2]. A number of metabolites present
in abnormal concentrations in these infants were therefore tested for
effects on globin expression. One metabolite, α amino-n-butyric acid
(ABA), was found to enhance γ globin and inhibit β globin expression
in cultured neonatal erythroid cells[3]. Further investigation now shows
that ABA, sodium butyrate and similar compounds can enhance γ globin

Molecular Biology of Erythropoiesis
Edited by J. L. Ascensao et al.
Plenum Press, New York

177

expression in cultured cells of patients with β globin disease, and that these agents can delay and inhibit the globin switch in an in vivo fetal animal model. The effects of butyric acid on histone acetylation may be a mechanism of action involved.

MATERIALS AND METHODS

Erythroid Progenitor Cultures

Peripheral blood was collected into preservative free heparin and the mononuclear cells were separated on Ficoll-Hypaque, washed, and cultured in IMDM methylcellulose medium with 2.5U/ml erythropoietin (Terry Fox Laboratories, Vancouver, Canada) 5U/mlGM-CSF (Amgen, Thousand Oaks, CA), with and without sodium butyrate and 10 agents consisting of slight modifications of butyric acid at concentrations of 0.1-0.5 mM, as previously described[4]. On day 12-13 of cell culture, Bfu-e were harvested and labelled with 25 μCi lyophilized ^{3}H-leucine in 250 μl of leucine-free minimum essential media with 15% fetal calf serum, and the same concentration of analogues in which the cells were cultured. Cultured erythroblast lysates were electrophoresed on acid-Triton-urea gels, and autoradiographs were prepared and scanned on a laser densitometer to quantitate proportions of α and non-α globin chains as previously described[4].

Histone analysis

Fetal liver cells were utilized in order to study a population of greater than 90% erythroid cells. A single cell suspension of 10^7 cells each were incubated at 37°C/5% CO_2 in Iscove's medium with 10% fetal calf serum alone and with 20 mM sodium butyrate, a halogenated compound designated 901409, L-α-amino-n-butyric acid, and D-chloro-alanine (designated D-594). After 48 hours of treatment, nuclei and acid extracts were prepared and the samples were dialyzed into 1M acetic acid. Protease inhibitors and sodium butyrate were added to all steps after incubation. 44 μg of the final extract was electrophoresed on acid-urea gels to resolve acetylated histones as previously described[5].

Fetal sheep studies

Venous and arterial catheters were placed in ovine fetuses by hysterotomy as previously described[6]. Butyric acid, neutralized to pH 7.4, and two analogues of butyric acid with minor molecular substitutions were infused into the fetus via a peristaltic pump. Blood was sampled daily and globin chain synthesis performed on ^{3}H-leucine labelled reticulocytes[6]. Results were compared with globin synthesis in ovine fetuses who were similarly operated on in utero, including some which were infused with normal saline at the same rate.

RESULTS

Cultures from twenty-five patients with sickle syndromes and β thalassemia were cultured with and without butryic acid compounds. In two-thirds, butyrate analogues increased γ globin expression by a mean of 12% over γ globin production found in control cultures from the same subject. Typical densitometry scans of autoradiograms from Bfu-e cultured from an infant with sickle cell anemia are shown in Figure 1. Shortly after birth, there was minimal γ globin expression in control cells (bottom panel). Addition of ABA increased γ globin expressed by 15%, and the effect was seen on both $^A\gamma$ and $^G\gamma$ globin genes (top panel). In most cases, the original $^G\gamma$:$^A\gamma$ ratios were not changed. The effects of ABA were rapid

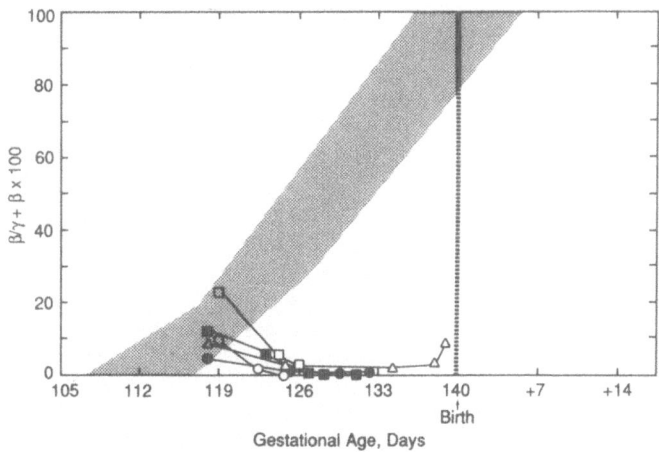

Fig. 1. Beta Globin Synthesis in Normal
and 907388-Treated Fetal Lambs.

Densitometric scans of globin produced by erythroid cultures from an
infant with sickle cell anemia. Controls cells procude only small residual
amounts of $^G\gamma$ and $^A\gamma$ globin (bottom
panel). Addition of α-amino-n-butyric
acid increased expression of both
globin genes (top panel).

and brief, occurring within six hours. Sodium butyrate produced a more
prolonged increase in γ globin synthesis, persisting throughout the culture
period. Compounds with minor substitutions of butyric acid were also
effective in increasing γ globin synthesis. In cultures from patients
with β thalassemia, in whom β globin was not detectable, the excess α
globin production found relative to γ globin in control cells was decreased
by 36% in the butyrate and ABA-treated cultures, indicating an increase
in γ globin expression.

In an attempt to confirm that butyric acid delays the globin gene
switch, an in vivo fetal ovine model was utilized. Infusions of sodium
butyrate into catherized ovine fetuses delayed the γ → β globin switch
in three of four fetuses. This effect was most prolonged when the infusion
were begun before β globin was greater than 10-15% of non-α globin.
In one fetus, β globin was low at the start of treatment; at term, after
three weeks' of butyrate therapy, there was no adult (β) globin detectable,
and this fetus produced 100% γ globin. When the infusions were begun
in one lamb with greater than 40% β globin production, the globin switch
was not delayed.

In an attempt to increase efficacy by prolonging the half-life of
butyrate, two compounds with halogen ion substitutions were infused into
fetal lambs. One compound caused profound acidosis. In fetuses infused
with a second analogue, designated 907388, a profound inhibition and
even reversal of adult globin production was found in all fetuses treated.
These results are illustrated in Figure 2.

Because butyrate has been shown to affect histone acetylation patterns
in other systems by inhibiting histone deacetylase, we examined whether
such an activity might be involved in its action on globin expression.

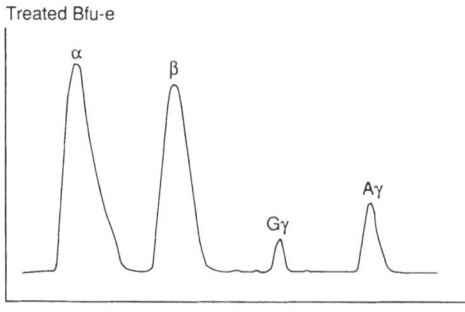

Treated Bfu-e

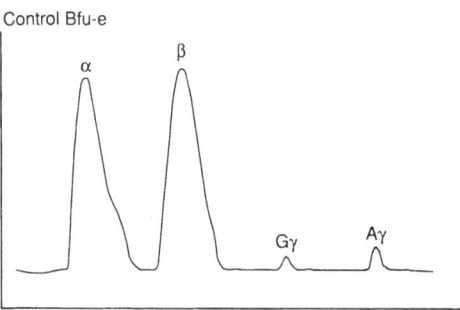

Control Bfu-e

Fig. 2.   Patient E.J. 3 1/2 Months.

The rise in β globin synthesis
in normal fetuses is shown
in the shaded area;   β globin
synthesis in fetuses treated
with the butyrate analogue
907338 in utero is shown by
the connected lines.   Adult
globin expression was profoundly inhibited.

We therefore compared the activity of various modified butyrate compounds
both for their activity in stimulating γ globin expression and for increasing
histone acetylation.   These results are illustrated in Figure 3.   The
degree of acetylation induced by the compounds correlated with their
degree of stimulation of γ globin synthesis.

DISCUSSION

These data demonstrate that a metabolite found in elevated concentrations in the plasma of infants of diabetic mothers, α amino-n-butyric
acid, and similar compounds, can increase γ globin and inhibit β globin
expression in erythroid cells in in vitro and in vivo experimental systems.
It is therefore likely that this metabolite causes the delayed globin
gene switch in these infants, interrupting a biologic clock.   Furthermore,
when added to the erythroid cells of patients with abnormal β globin
genes, an increase in γ globin production occurs even in short term cultures.
The amount of enhancement found, 12-30% over control levels, is a degree
that could be useful therapeutically.   This is extrapolated from evidence
of the selective survival of cells containing both Hb S and HbF, where
only 8% HbF synthesis in reticulocytes can produce >20% HbF in the peripheral
blood [7-10].

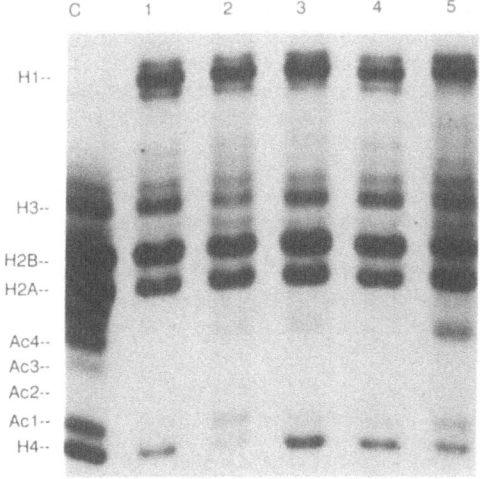

Fig. 3. Bulk histone acetylation
of fetal liver (erythroid)
cells treated with various
analogues of butyric acid.
Acetylated forms of histone
H4 are indicated on the left.
The degree of acetylation cor-
relates with efficacy of increasing γ globin expression
901409>sodium butyrate>D594>
L-α-amino-n-butyric acid. Untreated fetal liver cells are
shown in lane 1; calf thymus
core histone control is shown
in lane C.

It is intriguing that the same compounds can inhibit and even reverse
the fetal to adult globin switch in fetal sheep, as the sheep fetal globin
gene is structurally somewhat different from the human fetal globin gene.
However, butyrate has been shown to reactivate an embryonic globin gene
in adult chickens when given with 5-azacytidine by Ginder and colleagues
and Constantoulakis et. al. have found ABA to increase γ globin synthesis
in adult baboons[11,12]. The action of these compounds may therefore be
on several processes relating to active chromatin structure pertinent
to fetal globin expression rather than stimulating the expression of
a single gene.

Because butyrate is known to inhibit histone deacetylase and to
increase the acetylation of histones in other systems[13,14], we considered
that this activity could be involved in the action of butyrate on γ globin
expression. We therefore compared the effects of these compounds on
histone acetylation with their effects on γ globin. Finding a direct
correlation between the two activities suggest that an effect on histone
acetylation could indeed be involved with the stimulation of γ globin
expression, although our analysis was of global histone patterns.

Butyrate has been reported to induce differentiation of several
types of cells[15,16]. Rapid differentiation of erythroid cells could
be a possible mechanism for stimulating γ globin relative to β globin

expression, as γ globin is produced first during erythroid colony maturation. We have not found objective evidence for a difference in differentiation between control and butyrate-treated colonies[4]. Furthermore, some analogues produced larger colonies, others produced slightly smaller colonies than did control cultures, suggesting that no single effect on growth occurred with these analogues.

In summary, these data indicate that the normally set γ — β globin gene switch can be inhibited and even reversed by butyric acid and similar compounds. These agents can enhance γ globin expression to a significant degree and offer the potential for actually preventing expression of abnormal β globin genes. Histone acetylation may be involved in their activity, although other mechanisms of action should also be investigated.

REFERENCES

1. Wood, W.G., C. Bunch, S. Kelly, Y. Gunn, and G. Breckon. 1985. Control of hemoglobin switching by a developmental clock? Nature 313: 320.
2. Perrine, S.P., M.F. Greene, and D.V. Faller. 1985. Delayed in the fetal globin switch in infants or diabetic mothers. N. Eng. J. Med. 312: 334.
3. Perrine, S.P., B.A. Miller, M.F. Greene, R.A. Cohen, N. Cook, C. Shackleton, and D.V. Faller. 1987. Butyric acid analogues augment γ globin gene expression in neonatal erythroid progenitors. Biochem. Biophys. Res. Commun. 148: 694.
4. Perrine, S.P., B.A. Miller, D.V. Faller, R.A. Cohen, E.P. Vichinsky, D. Hurst, B.H. Lubin, and Th. Papayannopoulou. 1989. Sodium butyrate enhances fetal globin gene expression in erythroid progenitors of patients with HbSS and β thalassemia. Blood 74: 454.
5. Swerdlow, P.S., and A. Varsavsky. 1983. Affinity of HMG17 for a mononucleosome is not influenced by the presence of ubiquitin-H2A semihistone but strongly depends on DNA fragment size. Nucleic Acids Res. 11: 387.
6. Perrine, S.P., A. Rudolph, D.V. Faller, C. Roman, R.A. Cohen, S-J. Chen, and Y.W. Kan. 1988. Butyrate infusion in the ovine fetus delay the biologic clock for globin gene switching. Proc. Natl. Acad. Sci. USA 85: 8540.
7. Wood, W.G., M.E. Pembrey, G. Serjeant, R.P. Perrine, and D.J. Weatherall. 1980. Haemoglobin F synthesis in sickle cell anaemia: A comparison of Saudi Arab cases with those of African origin. Br. J. Haematol. 45: 431.
8. Reed, L.J., T.B. Bradley Jr., and H.M. Ranney. 1965. The effect of amelioration of anemia by the synthesis of fetal hemoglobin in sickle cell anemia. Blood 25: 37.
9. Nagel, R.L., R.M. Bookchin, J. Johnson, D. Labie, H. Wajcman, W.A. Isaac-Sodeye, G.R. Honig, G. Schiliro, J.H. Crookston, and K. Matsutomo. 1979. Structural bases of the inhibitory effects of hemoglobin F and A_2 on the polymerization of hemoglobin S. Proc. Natl. Acad. Sci. USA 75: 670.
10. Dover, G.J., S.H. Boyer, S. Charache, and K. Heintzelman. 1978. Individual variation in the production and survival of F cells in sickle cell disease. N. Engl. J. Med. 299: 1428.
11. Burns, L.J., J. Glauber, and G.D. Ginder. 1984. Butyrate induces selective gene transcription in injected Xeonopus oocytes: Enhancement by sodium butyrate. Embo J. 3: 2787.
12. Constantoulakis, P., Th. Papayannopoulou, and G. Stamatoyannopoulos. 1988. α-Amino-n-butyric acid stimulates fetal hemoglobin in the adult. Blood 22: 1961.

13. Riggs, M.G., R.G. Wittaker, J.R. Neuman, and V.W. Ingram. 1977. N-Butyrate causes histone modification in HeLa and Friend erythro-leukemia cells. Nature 268: 462.

14. Kruh, J. 1982. Effects of sodium butyrate, a new pharmacologic agent on cells in culture. Molec. Cell Biochem. 42: 65.

15. Darzynkiewicz, Z., F. Traganos, S-B. Xue, and M.R. Melamed. 1981. Effect of n-butyrate on cell cycle progression and in situ chromatin structure of L1210 cells. Exp. Cell. Res. 136: 279.

16. Prasad, K.N. 1980. Butyric acid, a small fatty acid with diverse biological functions. Life Sciences 27: 1351.

TUMOR NECROSIS FACTOR α AND THE ANEMIA OF CHRONIC DISEASE:

EFFECTS OF CHRONIC EXPOSURE TO TNF ON ERYTHROPOIESIS IN VIVO

G.D. Roodman, R.A. Johnson*, and U. Clibon

Research Service and Geriatric Research, Education,
  and Clinical Center
Veterans Administration Hospital
*Wilford Hall USAF Medical Center
San Antonio, Texas 78284

ABSTRACT

The anemia of chronic disease is associated with conditions in which
macrophage activation occurs. Activated marrow macrophages suppress
erythropoiesis in vitro, and produce tumor necrosis factor (TNF). Therefore,
we tested the effects of chronic in vivo exposure to TNF to determine
if it were a candidate for a mediator of the anemia of chronic disease.
Nude mice were inoculated with CHO cells expressing the human TNF gene,
or with control cells containing the transfection vector alone. The
TNF mice promptly became reticulocytopenic, and after 3 weeks their corrected reticulocytes were 2.6±.7% compared to 7.3±4% in control mice. The
hematocrit at 3 weeks was 28.4±1.7% in TNF mice compared to 46±8% in
control mice. This anemia was also associated with low serum iron and
normal iron stores and increased erythropoietin levels. The TNF mice
showed an absolute monocytosis with twice the number of circulating monocytes
as control mice. The TNF mice also became mildly thrombocytopenic.
Marrow CFU-E and BFU-E were profoundly decreased ($1.2\pm2\times10^3$ vs $8.6\pm2\times10^4$
CFU-E per femur, and $6.5\pm1\times10^2$ vs $8.5\pm.2\times10^4$ BFU-E per femur). Splenic
CFU-E and BFU-E were similarly depressed. In contrast, marrow CFU-GM
and CFU-GEMM were not affected. Administration of recombinant human
erythropoietin to these mice failed to reverse the suppressed erythropoiesis
in mice bearing TNF producing tumors. These data demonstrated that TNF
preferentially inhibits erythropoiesis in vivo, and may be important
in the pathogenesis of the anemia of chronic disease.

INTRODUCTION

The anemia of chronic disease (ACD) is seen in patients with chronic
infections, rheumatic diseases, and malignancies. Typically, the anemia
is hypoproliferative, nonprogressive, and associated with low serum iron
in the face of increased iron stores[1]. Although the clinical features
of ACD have been well characterized, the pathogenesis of ACD is unclear.
Impaired iron utilization has been implicated[2] but other studies have
found no consistent effects on iron metabolism[3]. High doses of intravenous
iron do not improve the anemia[4]. Insufficient erythropoietin production

Molecular Biology of Erythropoiesis
Edited by J. L. Ascensao et al.
Plenum Press, New York

185

has been found in some[5] but not all[6] studies. Shortened red blood cell survival has also been noted[7].

Recent studies of ACD have focused on the modulation of erythropoiesis by either immune products or by marrow macrophages. For example, interferons (alpha, beta, and gamma) suppress erythropoiesis in vitro[8]. The marrow macrophage may have a key role in the inhibiting erythropoiesis of ACD. Although resting macrophages, in physiologic numbered, enhance erythroid colony formation, increased numbers of macrophages markedly inhibit CFU-E and BFU-E growth[9]. Activated bone marrow macrophages from patients with ACD or with chronic fungal infections suppress erythropoiesis in vitro, and the suppression resolves when the underlying fungal infection is successfully treated[10,11]. Activated macrophages produce a number of cytokines which may affect erythropoiesis[12] including interleukin-1, interferons alpha and gamma, and tumor necrosis factor (TNF). Of these, TNF is a potential candidate as a macrophage mediator of ACD. Resting macrophages enhance erythroid colony formation[10] and do not produce TNF, but activation of macrophages by endotoxin produces a prompt increase in the expression of the TNF gene[13]. Membrane bound TNF accounts for most or all macrophage cytotoxic activity, and secreted TNF may be 1 to 2% of the cells total secretory product[14,15]. TNF suppresses murine and human erythroid progenitors in vitro[16-18] and repeated injections of TNF produce anemia in mice[19] and in humans[20]. However, intermittent injections of TNF may not be an appropriate physiologic model for conditions associated with chronic TNF production, such as may occur in ACD. Therefore, we tested the effect of chronic TNF exposure on erythropoiesis. A Chinese hamster ovary cell line which was transfected with the human TNF gene and constitutively produced high concentrations of TNF after transplantation into nude mice was used as a model system of chronic TNF exposure in vivo[21]. TNF induced a hypoferremic, hypoproliferative anemia with decreased numbers of marrow and splenic CFU-E and BFU-E but with little change in CFU-GM or CFU-GEMM. Erythropoietin failed to reverse the anemia induced by TNF. These data demonstrated that chronic TNF exposure in vivo results in preferential suppression of erythropoiesis.

METHODS

Cytokines

Recombinant human TNF and antibody to TNF were kindly supplied by Dr. Leo Linn and Dr. Abla Creasey, Cetus Corporation. TNF was assayed using the L929 cell model in the presence of actinomycin D[22]. Serum samples were frozen at $-70°C$ after collection until tested for the presence of TNF.

Cells

Chinese hamster ovary cells which had been transfected with the CMV-Neo transmission vector alone (CHO-cells) or with the vector containing the human TNF gene (TNF-cells)[21] were cultured in DMEM (Gibco) with 10% fetal calf serum (Hyclone) and 0.1% penicillin/streptomycin (Sigma) at $37°C$ in a humid atmosphere of 4% CO_2-air. The cells were harvested by trypsinization, and nude mice were inoculated in the right thigh with 10^7 cells in 0.5 ml of phosphate buffered saline pH 7.4 intramuscularly, using a 1cc syringe and 25 gauge needle. Tumor size was estimated by measuring and multiplying the two largest perpendicular diameters.

Mice

Male or female Balb/c nude mice (Audie Murphy VAH breeding colony) 6 to 8 weeks of age, were housed in laminar flow isolator cages with a 12 hour light and 12 hour dark cycle. In some experiments metabolic cages (Lab Products) were used. Water was acidified to pH 3 and supplemented with 0.5 cc multivitamins for infusion (Lymphomed) for every 2 leters. Autoclaved mouse chow (Purina) was provided ad lib. Mice were handled with glove and mask precautions. Mice inoculated with TNF cells will be referred to as "TNF mice" and mice inoculated with CHO-cells will be referred to as "control mice".

Blood Tests

Weekly phlebotomy was performed under metofane anesthesia and approximately 150 microliters of blood were obtained by puncturing the orbital plexus with heparinized capillary tubes. White blood cell and platelet counts were performed manually in a hematocytometer. Reticulocyte counts, spun hematocrit and 100 cell white blood cell differentials were also performed manually. The reticulocyte index was calculated by correcting the reticulocyte count to a hematocrit of 50%. Serum iron, iron binding capacity, albumin, triglycerides, and creatinine were determined on pooled specimens by multichannel autoanalyzer (Hitachi Model 736, Boehringer-Mannheim Diagnostics). Serum erythropoietin levels on pooled specimens were determined by RIA[23].

Progenitor Cell Assays

Marrow CFU-GM and CFU-GEMM and marrow and spleen CFU-E and BFU-E were assayed on individual mice. Bone marrow cells were flushed from femora using a 25 gauge needle and alpha MEM (Gibco). BFU-E and CFU-E were plated using a modification of the method of Iscove et al[24]. Quadruplicate 0.1 ml cultures containing $2x10^4$ cells in 10% charcoal-treated fetal calf serum (FCS) (Hyclone), 1% bovine serum albumin, 0.04 ml 2% methylcellulose (Fisher), 0.1 units human erythropoietin (Toyobo) and $5x10^{-5}M$ 2-mercaptoethanol were cultured in 96 well plates. Spleen cells suspensions were obtained by passing individual spleens through a stainless steel tissue sieve with alpha-MEM, and then pipetting the resultant suspension up and down with a 10ml pipette several times. Larger particles were allowed to settle for 2-3 minutes, and the remaining cells were resuspended and cultured for CFU-E and BFU-E as above, except that $4x10^4$ cells per well were plated. CFU-GM and CFU-GEMM assays were performed essentially as described by Nakahata and Ogawa[25]. Triplicate 1 ml cultures containing $2x10^5$ cells, 30% FCS, 5% pokeweed mitogen spleen cell conditioned media (PWM-SCCM)[26], 1 unit erythropoietin, $5x10^{-5}M$ 2-mercaptoethanol and 0.5 ml 2% methylcellulose were plated in 35 mm tissue culture dishes. Cultures were incubated at $37°C$ in a humid atmosphere of 4% CO_2-air. After 2 days of culture CFU-E were scored using an inverted microscope fitted with a Soret band filter. Wright staining of plucked colonies confirmed the accuracy of this method. After 6 or 7 days, BFU-E, CFU-GM and CFU-GEMM were scored in the same manner. CFU-E were clusters of 8 or more erythroid cells on day 2, BFU-E were colonies of 40 or more erythroid cells or composed of 2 or 3 subcolonies. Myeloid colonies of greater than 40 cells were scored as CFU-GM. Colonies with both myeloid and erythroid elements were scored as CFU-GEMM. Wright staining of plucked mixed colonies confirmed that these colonies contained erythroid, granulocytic, macrophage and megakaryocytic elements. Megakaryocyte progenitors (CFU-Meg) were estimated by plating triplicate 0.5 ml plasma clots, each containing $1x10^5$ marrow cells, 30% charcoal treated fetal calf serum, 5% PWM-SCCM, 0.5 units erythropoietin, $5x10^{-5}$ M β-mercaptoethanol, and 0.125 ml each of 4x thrombin (Parke-Davis) and citrated bovine plasma (Colorado Serum Company). Clots were harvested after 7 days, and megakaryocyte progenitors determined by staining for acetylcholinesterase activity[27,28].

Histology

Bone marrow cellularity and iron stores were assessed on decalcified vertebral columns, using hematoxylin and eosin, and Prussian-blue stains respectively.

Erythropoietin Treatment groups

Eight groups of mice were used for these studies. Group 1 was injected with CHO-control cells and treated with 1,000 IU rhEp/kg. Group 2 was not injected with any tumor cells and received 1,000 U rhEp/kg. Group 3 was injected with CHO-control cells and treated with saline carrier. Group 4 was injected with CHO-TNF cells and treated with saline carrier. Group 5, 6, 7, and 8 were injected with CHO-TNF cells and treated with 100, 250, 500 or 1,000 Iu/kg of rhEp respectively 3 times per week for three weeks. All groups contained 10 mice/group except 5, 6, and 7 which had 4 mice/group.

Statistical analysis

Results were analyzed by a two-way analysis of variance for repeated measures followed by Newman-Keules range test.

RESULTS

All TNF and control mice developed tumors. After one week, TNF mice were notably lethargic, and over the 3 weeks after inoculation, showed slightly deceased food consumption (26.3 ± 1.6 vs $32.4 \pm .3$ gm/mouse/week for control mice). Water consumption was modestly decreased (32.3 ± 1.7 vs 36.4 ± 1.4 ml/mouse/week for control mice). The control mice showed a steady weight gain, as expected for young mice. The TNF mice developed progressive weight loss and at 3 weeks had lost an average of 10% body weight compared to a 10% weight gain in control mice ($p < .05$). At sacrifice, the TNF mice had smaller tumors (48 ± 11 mm^2 vs 294 ± 82 mm^2 for control mice, $p < .05$), absent suprascapular fat pads, and had significantly elevated serum TNF levels compared to very low levels in control mice ($4.6°3.5 \times 10^{-10}$M vs $4.1 \pm 2 \times 10^{-12}$M, $p < 0.01$). The TNF level detected in control mice was at the limit of sensitivity for the TNF assay.

Mice bearing TNF producing tumors demonstrated a progressive decrease in their hematocrits to $28.4 \pm 1.7\%$ compared to $46 \pm 0.8\%$ in control mice ($p < .02$). Similar results were seen in 3 other experiments. This anemia was associated with a profound reticulocytopenia which was most severe after 1 week, and although the reticulocyte index increased thereafter, the reticulocyte response was inadequate for the degree of anemia. This nadir for reticulocyte index at week 1 in TNF mice was also observed in 4 separate experiments. TNF treated mice had much higher erythropoietin levels than controls or untreated mice. The erythropoietin level in pooled sera from TNF mice was 440 mU/ml compared with 66 mU/ml in control mice. For comparison untreated mice had a level of 31 mU/ml, and 2 days after phlebotomy to a hematocrit of 30%, the erythropoietin level rose to 160 mU/ml.

The serum iron level was decreased in TNF mice compared with controls (97 vs 248 micrograms per deciliter). These mice were not iron deficient, since marrow iron stores, assessed by Prussian-blue staining of vertebral marrow, were present in 10/10 TNF treated mice and in 7/10 control mice.

There was no consistent difference in white blood cell count between TNF and control mice, and in no experiment was leukopenia noted. There was, however, a consistent absolute monocytosis in mice bearing TNF producing tumors compared to control mice (absolute monocyte count in TNF mice = 5500±1400 vs 2030±650 x 10^9/L in control mice, p <0.02).

Platelet counts were modestly decreased in TNF mice compared to control mice, (650±40 vs 1,100±110 x 10^9/L, at 3, p=<.005) but remained near the normal range. Both TNF and control mice had abundant megakaryocytes on histologic examination of vertebral marrow.

TNF mice showed erythroid hypoplasia both by marrow histology and on Wright stained cytospin preparation of marrow cells (myeloid/erythroid ratios 46.4±15 for TNF mice vs 6.5±9 for control mice). Overall vertebral marrow cellularity was 80-100% in all mice.

Marrow cell recovery was decreased six-fold in TNF mice compared to control mice in all experiments (1.9±6 vs 11±0.9 x 10^6 per femur in TNF and control mice) despite similar vertebral marrow cellularity. Both marrow total CFU-E and BFU-E were profoundly decreased at 3 weeks in mice with TNF producing tumors (Table 1). In contrast, CFU-GEMM, CFU-GM, and CFU-Meg were not significantly changed. These differences were maintained whether results are expressed per 10^5 plated cells or per femur. Spleen cell recovery and spenic CFU-E and BFU-E were also decreased in TNF mice, compared to controls (Table 2). BFU-E derived colonies from TNF mice were clearly and consistently much smaller than colonies from control mice.

Table 1.   Marrow Progenitors In TNF And Control Mice

| | CONTROL | | TNF | |
	PER 10^5 Cells	PER FEMUR	PER 10^5 Cells	PER FEMUR
Cells	–	6.8±2 $x10^6$	–	2.7±.5 $x10^6$ *
CFU-E	770±200	8.5±.2 $x10^4$	68±13*	1.2±.2 $x10^3$ *
BFU-E	180±50	2.0±.5 $x10^4$	35±7*	6.5±1 $x10^2$ *
CFU-GEMM	30±5	1.9±.5 $x10^3$	43±12	1.3±.4 $x10^3$
CFU-GM	60±9	2.4±1 $x10^3$	60±20	1.8±.7 $x10^3$
CFU-Meg	16±3.6	1.1±2 $x10^3$	40±16	1.1±4 $x10^3$

Mice were inoculated with 10^7 CHO/TNF (TNF) or CHO/CMV-Neo (control) cells. After 21 days marrow cells were harvested and cultured for hematopoietic progenitors as described in Methods. Results are reported as mean ± SEM. The results per femur were calculated using the colony count per 10^5 nucleated cells and the number of nucleated cells per femur for individual mice, and the weighted mean is reported. For CFU-E and BFU-E n=10 mice per group, for CFU-GM, CFU-GEMM, and CFU-Meg n=5 mice per group.
* p<.01 for TNF compared to control.

Table 2. Splenic Progenitors In TNF And Control Mice

	Control (per spleen)	TNF (per spleen)	
Nucleated Cells ($\times 10^7$)	12.9±1.1	4.0±.8	*
CFU-E ($\times 10^4$)	5.7±2	1.6±.5	*
BFU-E	7400±860	372±80	*

Mice were inoculated with 10^7 CHO/TNF (TNF) or CHO/CMV-Neo
(control) cells. After 21 days spleens were harvested and
cultured for CFU-E and BFU-E as described in Methods. Results
are reported as mean ± SEM. n= 5 mice per group.

* p <.05 compared to control

The effects of erythropoietin treatment on the anemia induced by
TNF was then examined. The hematocrits in all animals bearing TNF producing
tumors treated with erythropoietin were not significantly different from
each other, regardless of the dose of rhEp given (Table 3). Mice not
bearing tumors and injected with rhEp had hematocrits of 60% or greater.
The hematocrits of TNF mice were significantly different (p<.02) from
control mice injected with saline or rhEp. Control mice treated with
rhEp had significantly higher hematocrits than control mice treated with
saline. Animals not bearing any tumor and treated with rhEp had significantly higher hematocrits than all other groups (p<.01) except control mice
treated with rhEp.

The results for CFU-E and BFU-E assays of marrow from TNF mice and
control mice treated with saline or rhEp are shown in Table 4. Only
the first 5 mice of each group had progenitor assays done. Marrow from
TNF mice treated with saline formed significantly fewer CFU-E and BFU-E
colonies compared to control mice treated with rhEp or saline. Treatment
of TNF-mice with rhEp increased CFU-E and BFU-E colony formation compared
to TNF-mice treated with saline, but this increase was not statistically
significant. Marrow from control mice treated with rhEp formed the greatest
number of CFU-E and BFU-E colonies, but because of the large variability
in colony formation this was not significantly different from control
mice treated with saline.

DISCUSSION

We have shown that the cytokine tumor necrosis factor, a product
of activated macrophages, is a potent inhibitor of erythropoiesis in
vivo, inducing a hypoproliferative, hypoferremic anemia with only modest
effects on leukocyte and platelet counts. In contrast, a large number
of reports show that TNF can suppress both erythroid and myeloid progenitors
in vitro[16-18, 29-31]. These reports suggest that factors other than
suppression of progenitors are involved in the hematologic effects of
TNF in vivo. TNF has been shown to induce several colony stimulating
factors, including G-CSF[32], GM-CSF[33,34], and M-CSF[35]. Munker and Koeffler
et al have shown that GM-CSF can "protect" CFU-GM from the suppressive

Table 3. Effects of Recombinant Human Erythropoietin on Hematocrits
of Mice Bearing TNF Producing Tumors.

Treatment	Hematocrits
No Tumor + rhEp (1000 u/Kg)	59.8±0.9
CHO-Cells + rhEp (1000 u/Kg)	58.6±2.0
TNF-Cells + rhEp (100 u/Kg)	30.4±0.9*
TNF-Cells + rhEp (250 u/Kg)	32.8±1.2*
TNF-Cells + rhEp (500 u/Kg)	33.6±1.7*
TNF-Cells + rhEp (1000 u/Kg)	34.2±1.2*
CHO-Cells + saline	47.0±1.5*
TNF-Cells + saline	30.4±0.9*

Results represent the mean ± SEM for each treament group (n=5).
Recombinant human erythropoietin (rhEp) was injected i.p. 3 times
per week for 3 weeks at the doses specified. Hematocrits were
done at day 22 after the first Ep injection.

*$p<.05$ compared to mice not bearing tumors treated with rhEp.

effects of TNF[36]. The combined actions of GM-CSF and M-CSF would explain
the lack of leukopenia and the monocytosis seen in TNF treated mice. Similarly, the lack of severe thrombocytopenia may reflect the action of
other hemopoietins on megakaryocyte progenitors. Our observation of
normal numbers of megakaryocytes and CFU-Meg are consistent with this
hypothesis.

The decrease in marrow cell recovery in TNF treated mice is unexplained.
Akahane et al found a similar decline after mice received several injections
of TNF[16]. We found no difference in histologic marrow cellularity between
TNF and control mice, and reticulin stains showed no evidence of significant
marrow fibrosis (data not shown). The decrease in total spleen cell
recovery is also unexplained, but both may be due to the cachectic state
of the mice.

TNF in vivo preferentially suppressed the erythroid progenitors
CFU-E and BFU-E, resulting in progressive anemia. The BFU-E derived
colonies from TNF mice were consistently fewer and smaller (estimated
40 cells) than BFU-E derived colonies from control mice (200-400 cells).
We have previously shown that human BFU-E derived colonies formed in
the presence of TNF are smaller and have a lower replating efficiency
than control colonies[10]. These data suggest that TNF has cytostatic
effects on the residual BFU-E. These data further suggest that the lower
numbers of CFU-E most probably represents decreased production of CFU-E
by a decreased BFU-E pool. In addition, these BFU-E have a lower proliferative capacity, further decreasing their ability to produce CFU-E.

Table 4. Effects of Erythropoietin on Erythroid Progenitor Growth in Marrow Cultures from Control Mice and TNF-Mice.

Treatment (n=5)	CFU-E/10^5 Cells	BFU-E/10^5 Cells
CHO-Cells + saline	120±19	20±2
CHO-Cells + rhEp (1000 u/Kg)	184±54	29±8
TNF-Cells + saline	50±7*	7±2*
TNF-Cells + rhEp (1000 u/Kg)	87+19*	14+2

*p<.05 compared to mice treated with CHO-cells + saline or rhEp.

Effects of recombinant human erythropoietin on erythroid progenitor cell growth in marrow cultures from contrpol mice and TNF-mice. Mice were inoculated with CHO-control cells or TNF-cells as described in Methods and treated with either saline (0.1 ml i.p. 3 times per week) for 3 weeks or with recombinant human erythropoietin (1000 per/Kg 3 times per week) for 3 weeks. At the end of the treatment the animals were sacrificed, and marrow cells were flushed from their femurs and cultured for CFU-E and BFU-E as described in Methods. There were 5 mice per group and all cultures were done in quadruplicate.

*p<.05 compared to animals treated with CHO-cells and saline or erythropoietin.

The low absolute numbers of erythroid progenitors may explain the progressive anemia seen in these animals. Lower concentrations of TNF might allow larger numbers of erythroid progenitors to persist and thus maintain a steady state level of erythropoiesis and produce a nonprogressive anemia in patients with ACD. However, because all our mice developed lethal tumors after 4-5 weeks, we could not demonstrate this.

Intermittent rhEp treatment failed to prevent or ameliorate the anemia produced in these mice continuously exposed to TNF. Although rhEp increased reticulocyte counts in TNF mice (data not shown), this increase was not sufficient to compensate for the suppressed erythropoiesis seen in these mice. Several lines of evidence suggest that this lack of response to rhEp is due to the cytotoxic effects of TNF on erythroid progenitors. Both marrow cellularity and CFU-E and BFU-E concentrations indicate that there was an inadequate number of erythroid progenitors available to respond to rhEp in mice bearing TNF producing tumors. Thus, although rhEp does increase the reticulocyte response initially in TNF mice, it cannot overcome the decreased number of available erythroid progenitors present in TNF mice.

An alternative possibility to explain the lack of responsiveness of TNF mice to rhEp is that the mice developed neutralizing antibodies to human erythropoietin. However, this seems unlikely. The increased hematocrits (>69%) were seen in control animals with no tumors treated with rhEp, demonstrating that the rhEp was active. In addition, we analyzed serum samples from all treatment groups after day 22 for immunoglobulins capable of binding erythropoietin and found none (data not shown). Thus, rhEp did not induce anti-Ep activity in our model system.

The data suggest that patients with very high circulating levels of TNF may not be able to respond to erythropoietin clinically. Further studies are required to determine if rhEp can overcome erythroid suppression seen with lower concentrations of TNF.

Our data and those of others suggest that TNF is a candidate for a mediator of anemia in patients with cancer and chronic inflammatory conditions: 1) Patients with malignancies and chronic inflammatory states typically have significantly elevated serum levels of TNF by radioimmunassay[37]. 2) Peripheral blood mononuclear cells from patients with solid tumors spontaneously produce large amounts of tumor necrosis factor[38]. If marrow macrophages in these patients also produced similar amounts of TNF, erythroid progenitors might be exposed to effectively higher concentrations of TNF than those found in serum. 3) Treatment of mice[16] and of humans in Phase I studies with TNF[20] results in anemia. 4) Our data show that the anemia produced in mice by TNF has several features in common with the anemia of chronic disease: It is hypoproliferative, with low serum iron in the face of normal iron stores, and is associated with appropriately elevated erythropoietin levels. In contrast to anemia of chronic disease (ACD), or mice had a progressive anemia. This may reflect the 100-1000 fold higher concentrations of TNF present in mice bearing TNF producing tumors compared to TNF levels reported in patients with conditions commonly associated with ACD[38]. In addition, ACD is by definition, a chronic condition, developing over months, while mice with TNF-producing tumors were anemic within 3 weeks. The rapidly growing tumors and high levels of TNF may have overcome the ability of resistant erythroid progenitors to maintain hematocrit levels.

In summary, chronic continuous in vivo exposure of nude mice to TNF produces preferential suppression of erythropoiesis which is not reversed by erythropoietin. TNF's suppression of erythropoiesis may be a direct or indirect effect of TNF. For example, TNF may induce production of interferon, a known suppressor of erythroid progenitors[8]. Alternatively erythropoiesis may be inhibited by a membrane bound activity on monocytes[39] which are increased in number in these mice by TNF. The degree of inhibition of erythropoiesis by TNF could also be related to the athymic state of these mice. However, the effect of TNF on erythropoiesis has not been shown to be strain or species specific[16-20]. Future models, such as using retroviral vectors to transfect genes into normal mice, may help clarify the role of TNF and other cytokines in the pathogenesis of the anemia of chronic disease.

ACKNOWLEDGMENTS

We wish to thank J. Laderer for secretarial assistance in typing the manuscript. This research was supported by Merit Review Funds from the Veterans Administration and grants AM-35188 from the National Institutes of Arthritis, Diabetes, Digestive and Kidney Diseases, HL-31264 from the Heart, Lung and Blood Institute, and CA40035 from the National Cancer Institute. Dr. Roodman is a recipient of a Clinical Investigator Award from the Veterans Administration Research Service.

REFERENCES

1. Lee, G.R. 1983. The anemia of chronic disease. <u>Seminars In Hematology</u> 20: 61.
2. Douglas, S.W., and J.W. Adamson. 1975. The anemia of chronic disorders: Studies of marrow regulation and iron metabolism. <u>Blood</u> 45: 55.
3. Zarrebi, M.H., R. Lysik, J. Distefano, and S. Zucker. 1977. The anemia of chronic disorders: Studies of iron reutilization in the anemia of experimental malignancy and chronic inflammation. <u>Br. J. Hemat.</u> 35: 647.
4. Kuhns, W.J., C.J. Gubler, G.E. Cartwright, and M.M. Wintrobe. 1950. The anemia of infection XIV, Response to massive doses of intravenously administered saccharated oxide of iron. <u>J. Clin. Invest.</u> 29: 1505.
5. Ward, W.P., J.E. Kurnick, and M.J. Pisarczyk. 1971. Serum level of erythropoietin in anemia associated with chronic infection, malignancy and primary hematopoietic disease. <u>J. Clin. Invest.</u> 50: 332.
6. Alexanian, R. 1972. Erythropoietin excretion of hemolytic anemia and in the hypoferremia of chronic disease. <u>Blood</u> 40: 946(a) (Abstr.).
7. Cartwright, G.E., and G.R. Lee. 1971. The anemia of chronic disorders. <u>Br. J. Hematol.</u> 21: 147.
8. Broxmeyer, H.E., L. Lu, E. Platzer, C. Feit, L. Juliano, and B.Y. Rubin. 1983. Comparative analysis of the influences of human gamma, alpha and beta interferons on human multipotential (CFU-GEMM), erythroid (BFU-E) and granulocytes (CFU-GM) progenitor cells. <u>J. Immunol.</u> 131: 1300.
9. Gordon, L.I., W.J. Miller, R.F. Branda, E.D. Zanjani, and H.S. Jacob. 1980. Regulation of erythroid colony formation by bone marrow macrophages. <u>Blood</u> 55: 1047.
10. Roodman, G.D., V.W. Horadam, and T.L. Wright. 1983. Inhibition of erythroid colony formation by autologous bone marrow adherent cell from patients with the anemia of chronic disease. <u>Blood</u> 62: 406.
11. Zanjani, E.D., P.B. McGlave, S.F. Davies, M. Banisadre, M.E. Kaplan, and G.S. Sarosi. 1982. In vitro suppression of erythropoiesis by bone marrow adherent cells from some patients with fungal infection. <u>Br. J. Hematol.</u> 50: 479.
12. Nathan, C.F. 1987. Secretory products of macrophages. <u>J. Clin. Invest.</u> 79: 319.
13. Collart, M.A., D. Belin, J.D. Vassalli, and P. Vassali. 1987. Modulations of functional activity in differentiated macrophages are accompanied by early and transient increase or decrease in c-fos gene transcription. <u>J. Immunol.</u> 139: 949.
14. Decker, T., M.L. Lohmann-Matthes, and G.E. Gifford. 1987. Cell associated tumor necrosis factor (TNF) as a killing mechanism of activated cytotoxic macrophages. <u>J. Immunol.</u> 138: 957.
15. Beutler, B., J. Mahoney, N.L. Trang, P. Pekela, and A. Cerami. 1985. Purification of cachetin, a lipoprotein-lipase suppressing hormone secreted by endotoxin induced RAW 264.7 cells. <u>J. Exp. Med.</u> 161: 984.
16. Akahane, K., T. Hosoi, A. Urabe, M. Kawakama, and F. Takaku. 1987. Effects of recombinant human tumor necrosis factor (rhTNF) on normal human and mouse hemopoietic progenitor cells. <u>Intl. J. Cell Cloning</u> 5: 16.
17. Roodman, G.D., A. Bird, D. Hutzler, and W. Montgomery. 1987. Tumor necrosis factor-α and hematopoietic progenitors: The effects of tumor necrosis factor on the growth of erythroid progenitors CFU-E and BFU-E and the hematopoietic cell lines K562, HL60, HEL cells. <u>Exp. Hematol.</u> 15: 928.

18. Murphy, M., G. Perussia, and G. Trinchieri. 1988. Effects of recombinant tumor necrosis factor, lymphotoxin, and immune interferon on proliferation and differentiation of enriched hemotopoietic precursor cells. Exp. Hematol. 16: 131.

19. Wei, H., K. Tracy, K. Manogue, N. Nguyen, Y. Fong, D. Hesse, B. Beutler, R. Solomon, A. Cerami, and S. Lowry. 1987. Cachetin mediates suppressed food intake and anemia during chronic administration. Federation Proceedings 46: 5963.

20. Blick, M., S.A. Sherwin, M. Rosenblum, and J. Gutterman. 1987. Phase I study of recombinant tumor necrosis factor in cancer patients. Cancer Res. 47: 2986.

21. Oliff, A., D. Defeo-Jones, M. Boyer, D. Martinez, D. Kiefer, G. Vuocolo, A. Wolfe, and S.H. Socher. 1987. Tumors secreting human TNF/cachectin induce cachexia in nude mice. Cell 50: 555.

22. Tsujimoto, M., Y.K. Yip, and J. Vilcek. 1985. Tumor necrosis factor: Specific binding and internalization in sensitive and resistant cells. Proc. Natl. Acad. Sci. USA. 82: 7626.

23. Birgegard G., O. Miller, J. Caro, and A. Ersler. 1982. Serum erythropoietin levels by radioimmunoassay in polycythemia. Scand. J. Haematol. 29: 161.

24. Iscove, N.N., F. Sieber, and K.H. Winterhalter. 1974. Erythroid colony formation in cultures of mouse and human bone marrow: Analyses of the requirement for erythropoietin by gel filtration and affinity chromotography on agarose-concavalin A. J. Cell Physiol. 83: 309.

25. Nakahata, T., and M. Ogawa. 1982. Clonal origin of murine hematopoietic colonies with apparent restriction to granulocyte--macrophage-megakaryocytes (GMM) differentiation. J. Cell Physiol. 111: 239.

26. Humphries, R.K., A.C. Eaves, and C.J. Eaves. 1979. Characterization of a primitive erythropoietic progenitor found in mouse marrow before and after several weeks in culture. Blood 53: 746.

27. Karnoviky, M.J., and L. Roots. 1964. A "direct-coloring" thiocholine method for cholinesterasese. J. Histochem Cytochem. 12: 219.

28. Jackson, C.W. 1973. Cholinesterase as a possible marker for early cells of the megakaryocytic series. Blood 42: 413.

29. Murase, T., T. Hotta, H. Saito, and R. Ohno. 1987. Effect of recombinant human tumor necrosis factor on the colony growth of human leukemia progenitor cells and normal hematopoietic progenitor cells. Blood 69: 467.

30. Abboud, S.L., S.L. Gerson, and N.A. Berger. 1987. The effect of tumor necrosis factor on normal human hematopoietic progenitors. Cancer 60: 2965.

31. Broxmeyer, H.E., D.E. Williams, L. Lu, S. Cooper, S.L. Anderson, G.S. Beyer, R. Hoffman, and B.Y. Rubin. 1986. The suppressive influences of human tumor necrosis factor on bone marrow hematopoietic progenitor cells from normal donors and patients with leukemia: Synergism of tumor necrosis factor and interferon γ. J. of Immunology 136: 4487.

32. Koeffler, H.P., J. Gasson, J. Raynard, L. Souza, N. Shepard, and R. Munker. 1987. Recombinant human TNF-α stimulates production of granulocyte colony-stimulating factor. Blood 70: 55.

33. Munker, R., J. Gasson, M. Ogawa, and H.P. Koeffler. 1986. Recombinant human TNF induces production of granulocyte-macrophage colony-stimulating factor. Nature 323: 79.

34. Broudy, V.C., K. Kaushansky, G.M. Segal, J.M. Harlan, and J.W. Adamson. 1986. Tumor necrosis factor α stimulates human endothelial cells to produce granulocyte/macrophage colony stimulating factor. Proc. Natl. Acad. Sci., USA 83: 7467.

35. Oster, W., A. Lindermann, S. Horn, R. Mertelsmann, and F. Herman.
       1987. Tumor necrosis factor (TNF)-alpha but not TNF-beta induces
       secretion of colony stimulating factor for macrophages (CSF-1)
       by human monocytes. Blood 70: 1700.
36. Munker, R., and P. Koeffler. 1987. In vitro action of tumor
       necrosis factor on myeloid leukemia cells. Blood 69: 1102.
37. Teppo, A.M., and C.P.J. Maury. 1987. Radioimmunoassay of tumor
       necrosis factor in serum. Clin. Chem. 33: 2024.
38. Aderka, D., S. Fisher, Y. Levo, H. Holtman, T. Hahn, and D. Wallach.
       1985. Cachetin/tumor-necrosis factor production by cancer
       patients. Lancet 2: 1190.
39. Feldman, L., C.M. Cohen, and N. Dainiak. 1986. In vitro release
       of physically separable factors from monocytes that exert opposing
       effects on erythropoiesis. Blood 67: 1454.

ERYTHROPOIESIS IN CANCER PATIENTS

UNDERGOING IMMUNOTHERAPY

Joao L. Ascensao[1], Shu-Jun Liu[2], Jaime Caro[3],
Eckhard Podack[4], Abraham Mittelman[5], Esmail D. Zanjani[6],
and Yu-Liang Zhao[2]

[1]Department of Medicine, Division of Hematology-Oncology,
Room L-3062, University of Connecticut Health Center,
Farmington, Ct., USA, [2]Beijing Institute for Cancer Research,
Beijing, People's Republic of China, [3]Cardeza Foundation for
Hematologic Research, Jefferson Medical College, Philadelphia,
PA, USA, [4]Comprehensive Cancer Center, University of Miami,
Miami, FL, USA, [5]Department of Medicine, Division of Neoplastic
Diseases, New York Medical College, Valhalla, NY, USA,
[6]University of Nevada School of Medicine, Reno, NV, USA

ABSTRACT

We studied ten patients with various types of cancer who were being
treated with Interleukin-2 (IL-2) and lymphokine activated killer cells
(LAK). All patients developed a reticulocytopenic, normochromic, normocytic
anemia. We noted some variability but no significant suppression of circulating erythroid progenitors. The levels of erythropoietin were lower than
expected for the hemoglobin/hematocrit values.

We could not detect Interferon or Tumor Necrosis Factor (TNF) in
the serum of these patients; however, the supernatant of LAK cells did
contain Interferon and TNF which could be neutralized with appropriate
antibodies.

These results suggest that the etiology of this anemia is multi-factorial.
Administration of recombinant erythropoietin (Ep) may be of benefit in
some of these patients.

INTRODUCTION

Erythropoietin (Ep) is a major regulator of erythropoiesis in vivo
and in vitro. The regulation of the production of this hormone has been
well described[1]. Altered levels of Ep can be found in some patients with
cancer[2] and, in some studies, in the anemia of chronic disease[2]. The
reason for this abnormally low level of Ep production is not fully clear.
Additionally, immunocompetent cells such as monocytes, lymphocytes and
NK cells can affect erythropoiesis in vitro[3]; helper and suppressor roles
have been ascribed to these cells. Cytokines produced by monocytes (TNF,
IL-1) can "shut off" erythropoiesis in vivo, strongly indicating a patho-
physiologic role for these factors in erythropoiesis. In aplastic anemia,

Molecular Biology of Erythropoiesis
Edited by J. L. Ascensao et al.
Plenum Press, New York

197

T-lymphocytes appear to play an important pathogenetic role[4]. In vitro mitogen activation of T-lymphocytes can lead to production of hemopoietic growth factors[5] as well as to generation of inhibitory cells[6] and factors[7].

Soluble factors produced by lymphocytes have been termed lymphokines; IL-2 supports the growth and proliferation of T-lymphocytes. Evidence indicates that IL-2 augments natural killer cell activity (NK) and the proliferation of cytotoxic lymphocytes[8].

In the present studies we used the in vivo model of administration of Interleukin-2 (IL-2) and lymphokine activated killer cells (LAK) to study the effects of activated immunocompetent cells on autologous hemopoiesis. Recently, LAK cells have been considered to have a potential therapeutic role in neoplastic disorders[9]. Patients receiving IL-2/LAK cells invariably develop reticulocytopenic anemia, eosinophilia, and, on occasion, thrombocytopenia[10]. It remains possible that hematologic alterations in these patients result from alterations of accessory cell function or from altered production responsiveness to appropriate hemopoietic growth factors. The present study was undertaken to more fully evaluate the regulatory effects of LAK cells on hematopoiesis.

MATERIALS AND METHODS

Patient samples

Heparinized blood (20 ml) and serum (8 ml) were collected on day of admission, at leukapheresis, the first day of administration of LAK cells, and then at weekly intervals. The blood was separated on a ficoll gradient, the mononuclear cells counted and used for culture as previously described[11]. The serum was stored at -70° C until used. Patients were given IL-2 on admission, leukapheresed twice daily and given IL-2 and LAK cells for approximately twelve days.

Bone marrow cells

Heparinized bone marrow was aspirated from normal volunteers and mononuclear cells prepared as previously described[11]. Monocyte depleted cells were obtained after removal of cells adherent to fetal calf serum (FCS) coated plastic flasks. T-cell depletion was accomplished by removal of cells which had rosetted with sheep red blood cells[11]. Less than 2% contaminating monocytes or T-cells were found in the depleted population

Hemopoietic stem cell assays

1. Erythropoiesis

(BFU-E/CFU-E): PBMNC or BMNC were cultured in plasma clot cultures as previously described[11]. Colonies and bursts were scored at days 7 and 14 respectively. A methylcellulose system was also used to study erythropoiesis in culture. The cells were cultured in triplicate with or without LAK cells or LAK-CM with erythropoietin (Ep) added at day 0 or at day 3. The plates were scored on day 14 and colonies were examined for numbers of constituents using a dissection microscope.

2. Myelopoiesis

(CFU-GM): Peripheral blood mononuclear cells (PBMNC) or nonadherent bone marrow mononuclear cells (BMNC) were cultured in triplicate 0.5 ml agar cultures with or without a source of colony stimulating factor (CSF) (GCT-CM, GIBCO) or media conditioned by LAK cells (LAK-CM). Colonies of more than 40 cells were scored on day 14 using a dissection microscope[11]. In addition, eosinophilic colonies (CFU-Eo) were evaluated in some cultures. Detection of CFU-Eo by morphology and staining characteristics has been previously described[12].

RESULTS

We have studied ten patients treated with our LAK/IL-2 protocol (Figure 1). All of the patients developed reticulocytopenic anemia (HCT < 35%, corrected reticulocyte < 1%); leukocytosis with eosinophilia (< 500/mm^3). Circulating erythroid progenitors (BFU-E) assessed in plasma clot cultures did not change appreciably; in one case, an increase in circulating BFU-E was seen during therapy (Table 1). The hemoglobinization and size of the erythroid bursts were similar to those of normal individuals.

We next examined the effect of LAK cells and LAK-CM on autologous BFU-E growth in the presence of Ep. Results from three experiments demonstrated that peripheral blood BFU-E formation was significantly inhibited by coincubation with LAK cells at titers of (1:1) (Figure 2). In no instance did BFU-E growth occur in the absence of Ep. Addition of similar titers of normal lymphocytes to the cultures had no inhibitory effect and in fact stimulated BFU-E growth (data not shown). Furthermore, inclusion of LAK-CM in the cultures inhibited BFU-E growth in a dose dependent manner (Figure 3). However, when Ep was added at day 3 (to test for BPA in the LAK-CM) there was evidence of this activity present. Erythroid colony formation (CFU-E) early in culture by LAK cells or donor's PBMC was undetectable even though morphological erythroid progenitors were present.

Since interferons may have modulatory effect on hematopoiesis, we therefore determined gamma interferon (gIF) levels in LAK-CM using a commercial radioimmuno assay (Centocor). Results from the RIAs revealed that LAK-CM contained varying levels of gIF (Table 2) and that the levels varied from patient to patient and appeared to decrease with time in _in vitro_ culture. At no time did the gIF levels exceed 500 U/ml.

In the next series of experiments, _in vitro_ myelopoiesis was examined using normal allogeneic bone marrow cells (BMNC) and conditioned media prepared from PBMC or LAK cells. Examination of PBMC for circulating

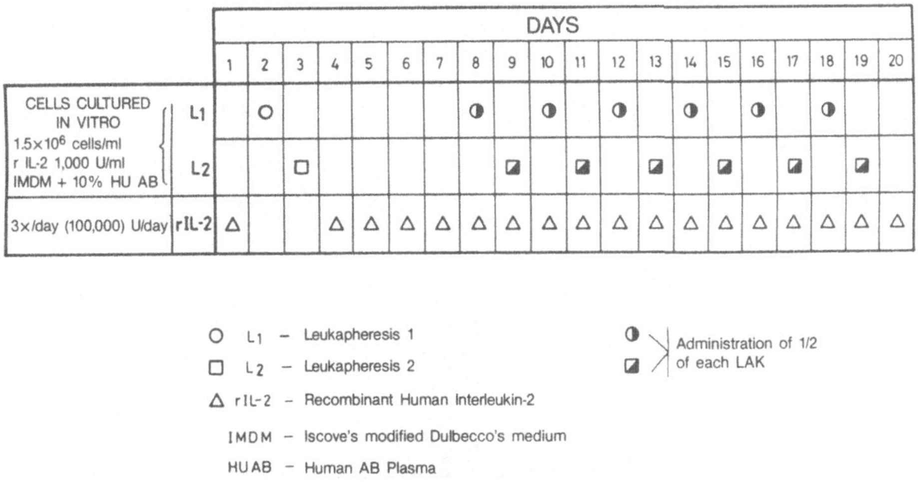

		DAYS																			
		1	2	3	4	5	6	7	8	9	10	11	12	13	14	15	16	17	18	19	20
CELLS CULTURED IN VITRO 1.5×10⁶ cells/ml r IL-2 1,000 U/ml IMDM + 10% HU AB	L1	O							◐		◐		◐		◐		◐	◐			
	L2			□					◪		◪		◪		◪		◪		◪		
3×/day (100,000) U/day	rIL-2	△			△	△	△	△	△	△	△	△	△	△	△	△	△	△	△	△	△

O   L₁ — Leukapheresis 1

□   L₂ — Leukapheresis 2

△   rIL-2 — Recombinant Human Interleukin-2

IMDM — Iscove's modified Dulbecco's medium

HUAB — Human AB Plasma

◐ ◪ Administration of 1/2 of each LAK

Fig. 1.   IL-2/LAK Protocol

Table 1. Circulating BFU-E: Effect of IL-2 and LAK/IL-2 BFU-E/Clot

PATIENT:	d0	d3	d8	d11	d18
BE	----	8.75+3.93	11.5+1.21	24.25+3.86	38.5+3.81
AS	8.75+1.64	8.25+2.34	4.25+1.75	7+2.70	4.70+1.41
EB	6.25+1.39	5.50+1.80	4.75+0.54	-----	-----
HA	19.75+1.14	19.25+2.07	15.00+2.07	-----	-----

myeloid progenitors revealed the presence of few if any colony forming cells (1-5 CFU/plate), which is similar to that obtained from normal individuals. Results revealed that when LAK-CM was included in normal BMNC agar cultures, without a standard source of CSF (GCT), myeloid colony growth was obtained. Furthermore, the colonies seen with LAK-CM contained more eosinophilic elements, whereas those seen with GCT (or standard CSF) were predominantly granulocytic (Figure 4).

Erythropoietin levels were determined using a radioimmunoassay. No Ep was detected in the LAK-CM. Table 3 depicts the correlation of Ep levels, reticulocyte counts and hematocrit in a representative patient. In general, the levels of Ep fluctuated above normal but were not appropriate for the degrees of anemia (Table 4). In some patients cessation of immunotherapy failed to restore production of Ep to appropriate levels for their anemia.

DISCUSSION

Patients with cancer treated with immunotherapy often develop anemia[10,13]. It is possible that this represents a worsening of the anemia of chronic disease seen as a subset of these patients or that it represents an altered response of their hemopoietic progenitors when subjected to a microenvironmental stress. We analyzed the effects of IL-2 activated LAK cells and their products on hemopoiesis. LAK cells and supernatants appeared to inhibit the growth of erythroid progenitors; some of this inhibition was due to interferon gamma present in the LAK-CM and possibly due also to TNF. However, at no time were we able to demonstrate interferon gamma or TNF in the serum of these patients. Furthermore, contrary to previously published results[13], we did not find any evidence of inhibition of circulating

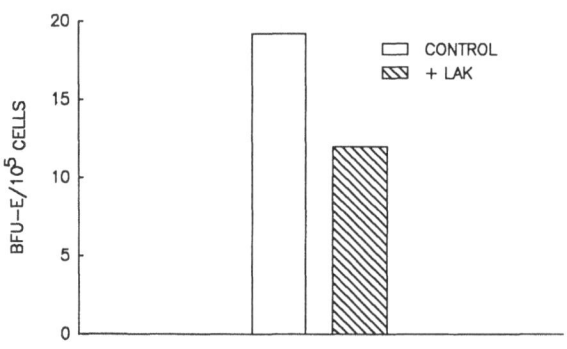

Fig. 2. Effect of LAK Cells on BFU-E

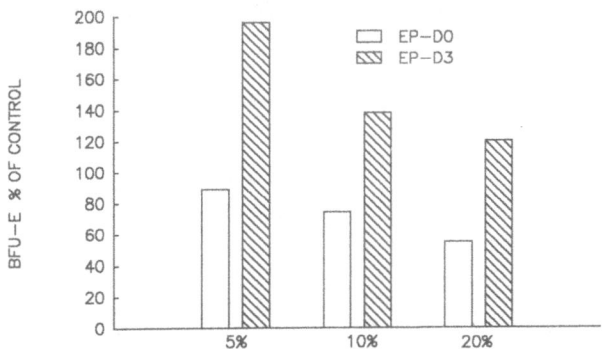

Fig. 3.   Effect of LAK-CM on Bone Marrow BFU-E

Table 2.   Interferon Gamma Determination by RIA in Supernatants of IL-2 Treated Cells IF-U/ML

PATIENT	d8	d11	d18
A.S.	210	223	150
B.E.	---	180	27
H.A.	6.8	9.2	9.2
E.B.	500	---	120

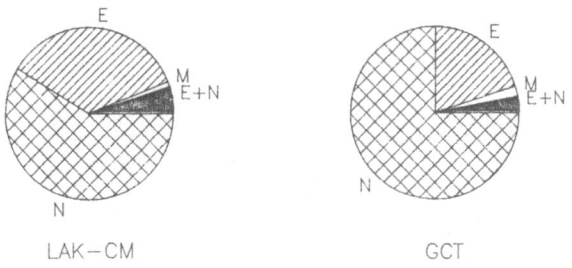

Fig. 4.

hemopoietic progenitors. This may reflect the differences in the protocols
used to treat these patients.

It seems reasonable for us to hypothesize that this altered hemopoiesis
is similar to the anemia of chronic disease. In this case, activated
immunocompetent cells elaborate cytokines which appear to selectively
inhibit erythroid progenitors. In this regard, Roodman et al, in a chapter
of this book[14], used an in vivo model which elegantly demonstrates that
excessive TNF production can cause anemia. Furthermore, it has been demonstrated that the interferons are more potent inhibitors of erythropoiesis
than myelopoiesis[15] and we and others have demonstrated an additive and

201

Table 3. Serum EP Levels in a Patient Treated With LAK/IL-2

DAY OF RX	EP LEVEL (mU/ml)	HCT	RETIC (%)
d0	36	40.7	----
d3	24	40	1.3
d8	84	31.3	2.4
d11	44	27.4	1.3
15	28	27.2	-----
19	90	26.7	2.5

Normal 16 mU/ml

P.V. < 10 mU/ml

Table 4. Composite of the Levels of EP in Several
Patients Receiving IL-2/LAK Cells

DAY	S.T.	K.U.	H.A.	K.E.	D.B.
0	56	21	16	23	----
3	21	30	27	28	----
8	25	45	17	140	48
4	13	35	----	---	36
14	13	50	20	28	28
16-18	13	38	----	50	24
24	24	----	----	----	----
30-33	30	----	28	----	----
60	----	----	----	----	13

Control 16 mU/ml

synergistic activity for the various cytokines[16]. NK cells are known
to produce TNF[17] as well as BPA[18]. The prolonged incubations of our LAK
cells in vitro, in the presence of IL-2, resulted in significant alterations
of the phenotype of these cells. The frequency of Leu 11 (CD 16) positive
NK cells increased from 5% (day 0) to 60% (day + 16) at the end of the
culture period. CD 8 paralleled Leu 11, whereas CD 3 and CD 4 decreased
in a reciprocal fashion. Cytotoxicity activity reached maximal activity

on days 12-14 and was 3-6 fold that of baseline. Thus it is possible
that not only did the IL-2 recruit cells in vivo, some of which have suppressive effects on erythropoiesis[19], but that the administration of a highly
enriched and activated population of cytotoxic cells may have contributed
to the suppression of erythroid progenitor development. This was not
reflected in a diminution of the circulating erythroid compartment, which
may have several explanations as we previously proposed[10]. Studies of
the marrow compartment would have been helpful but were deemed unnecessary
or too risky in these very ill individuals. Certainly, it was clear that
not only IL-2 but also LAK cells could stimulate eosinophilia. There
is substantial in vitro data on which to base these results[20].

The lack of erythropoietin response in these patients may be difficult
to explain, but may reflect vascular toxicity induced by the immunocompetent
cells and their soluble products. There may have been an underlying process
which contributed to the lack of Ep response, since in several patients
the levels of Ep remain low for some time after discontinuation of therapy.

It does appear that a multifactorial origin exists for this type
of anemia. The therapeutic strategy should be discussed in light of the
results obtained and may include the in vivo stimulation of erythropoiesis
by Ep.

REFERENCES

1. Zanjani, E.D., and J.L. Ascensao. 1989. Erythropoietin. Transfusion
        29:46.
2. Erslev, A.J., J. Wilson, and J. Caro. 1987. Erythropoietin titers
        in anemic, nonuremic patients. J. Lab. Clin. Med. 109: 429.
3. Ascensao, J.L., and E.D. Zanjani. (in press). Cellular interactions
        in hemopoiesis. In "Handbook of Hemopoietic Microenvironment"
        M. Tavassoli, Ed., Humana Press, NY.
4. Ascensao, J.L., R. Pahwa, and E.A. Kagan, et al. 1976. Aplastic
        anemia: evidence for an immunologic mechanism. Lancet 1: 669.
5. Meytes, D., A. Ma, and J.A. Ortega, et al. 1979. Human erythroid
        burst-promoting activity produced by phytohemoglutinin-stimulated
        radioresistant peripheral blood mononuclear cells. Blood 54:
        1050.
6. Banisadre, M., R.C. Ash, and J.L. Ascensao, et al. 1981. Suppression
        of erythropoiesis by mitogen-activated T-lymphocytes in vitro.
        In "Experimental Hematology Today". S.J. Baum, G.D. Gedney,
        A. Kahn, eds. Karger, NY pp 151.
7. Zoumbos, N.C., J.Y. Djeu, and N.S. Young. 1984. Interferon is the
        suppressor of hematopoiesis generated by stimulating lymphocytes
        in vitro. J. Immunol. 133:769.
8. Smith, K.A. 1980. T-cell growth factor. Immunol. Rev. 51:337.
9. Rosenberg, S.A., M.T. Lotze, and L.M. Muul, et al. 1985. Observa-
        tions on the systemic administration of autologous lymphokine-
        activated killer cell and recombinant interleukin-2 to patients
        with metastatic cancer. N. Engl. J. Med. 313: 1485.
10. Liu, S-J., J.L. Ascensao, and E. Podack, et al. 1987. Cellular
        interactions on hemopoiesis. Blood Cells 13: 101.
11. Ascensao, J.L., G.M. Vercellotti, and H.S. Jacobs, et al. 1984.
        Role of endothelial cells in human hematopoiesis: modulation
        of mixed colony growth in vitro. Blood 63: 553.
12. Phillips, P.G., G. Chikappa, and P.S. Brinson. 1983. A Triple
        staining technique to evaluate monocyte neutrophil and eosinophil
        proliferation in soft agar cultures. Exp. Hematol. 11: 10.
13. Ettinghausen, S.E., J.G. Moore, and D.E. White, et al. 1987. Hema-
        tological effects of immunotherapy with lymphokine activated

killer cells and recombinant interleukin-2 in cancer patients. Blood 69: 1654.

14. Roodman, G.D., R.A. Johnson, and U. Glibon. (in press). Tumor necrosis factor and the anemia of chronic disease: effect of chronic exposure to TNF on erythropoiesis in vivo In: "Molecular Biology of Erythropoiesis" J.L. Ascensao, et al, (Eds), Plenum Press.

15. Mamus, S.W., S. Beck-Schroder, and E. Zanjani. 1985. Supression of normal human erythropoiesis by gamma interferon in vitro: role of monocytes and T-cells. J. Clin. Invest. 75: 1496.

16. Zoumbos, N., E. Raefsky, and N. Young. 1986. Lymphokines and hematopoiesis. Prog. Hemat. 14: 201.

17. Degliantoni, G., M. Murphy, and M. Kobayashi, et al. 1985. Natural killer (NK) cell-derived hematopoietic colony-inhibiting activity and NK cytotoxic factor. J. Exp. Med. 162: 1512.

18. Pistoia, V., R. Ghio, and A. Nocera, et al. 1985. Large granular lymphocytes have a promoting activity on human peripheral blood erythroid burst forming units. Blood 65: 464.

19. Burdach, S.E.G., and L.J. Levitt. 1987. Receptor-specific inhibition of bone marrow erythropoiesis by recombinant DNA-derived interleukin-2. Blood 69: 1368.

20. Azuma, C., T. Tanabe, and M. Konishi, et al. 1986. Cloning of cDNA for human T-cell replacing factor (IL-5) and comparison with the murine homologue. Nucleic Acid Res. 14: 9149.

ERYTHROPOIESIS IN MURINE LONG TERM MARROW CULTURES

Mitsuo Konno, Cheryl Hardy and Mehdi Tavassoli

Division of Hematology-Oncology
V.A. Medical Center
Jackson, Mississippi 39216

ABSTRACT

We utilized a system of long-term bone marrow cultures that favored erythropoiesis and allowed the production of CFU-E for four weeks. Certain conditions of culture were crucial for erythropoiesis to occur. These conditions have been discussed. This system should allow further in vitro study of the regulation of erythropoiesis.

INTRODUCTION

Within the mammalian bone marrow, at least two elements interact to bring about hemopoiesis-the stem cell and stroma. Structurally, the marrow is organized within the bone so that the stem cell is permitted to interact not only with stromal cells but also with positive and negative regulatory molecules of the marrow microenvironment viz. extracellular matrix components and families of molecules such as the colony stimulating factors, interleukins, and hormones[1-7].

The net result of these interactions is two-fold: (a) self-renewal of stem cells in order to retain the size of the stem cell pool, despite continuous drainage to form mature blood cells; and (b) commitment of stem cells to various differentiation pathways, the most noted being myeloid, erythroid or megakaryocytic pathways.

Long-Term Bone Marrow Cultures (LTBMC)

This interaction of stem cell and stroma can be duplicated in vitro by a technique known as long-term bone marrow culture. The technique was first introduced by Dexter et al., and subsequently improved by Greenberger and others[8-10]. The technique exploited the fact that stromal cells of bone marrow are adherent to glass and other support matrices. When fully adherent stroma is developed over such matrices, it can then support the growth, proliferation and differentiation of hemopoietic cells. Mature cells are then released into the supernate where they can be recovered every week and, thus, the number of cells produced in the system per week can be determined. In addition to mature cells, progenitor cells are also produced in this system and are in a state of equilibrium between the supernate and the adherent layer. Thus, one

Molecular Biology of Erythropoiesis
Edited by J. L. Ascensao et al.
Plenum Press, New York

205

can measure, by specific assays, multipotential stem cells (CFU-S or CFU-Mix), bipotential progenitors (CFU-GM) or unipotent progenitors (BFU-E, CFU-E).

Standard LTBMC have served extensively to study the kinetics of stromal-progenitor cell relationships in granulopoiesis. This is because these standard cultures produced both mature cells and progenitors of granulocytes in ample number. Such a standard technique does not lend itself to the study of erythropoiesis and megakaryopoiesis: erythropoiesis does not proceed beyond BFU-E[11], and only a few megakaryocyte progenitors are produced in these cultures[12].

The purpose of this communication is to describe technical alterations that may permit the application of LTBMC to erythropoiesis. We first describe attempts by other investigators in this area, and then shall add our own experience to theirs. Finally, we describe different modifications of the standard technique required for application of LTBMC to erythropoiesis.

Erythropoiesis in LTBMC

In standard LTBMC relatively mature erythroid progenitors, day-2 CFU-E as well as morphologically recognizable mature erythrocytes are absent. Eliason et al.[13] found that by creating a culture environment less favorable to attachment of late erythroid progenitors, a progression through erythroid maturation could be observed. This can be done by the use of mechanical agitation or other manipulation. They were able to detect, in the presence of erythropoietin, late BFU-E, day-2 CFU-E and hemoglobinized, nucleated and non-nucleated red cells. Subsequently, Dexter et al.[14] showed that the addition of anemic mouse serum or normal mouse serum plus erythropoietin resulted in erythroid cell development, circumventing the need for mechanical agitation. Presumably a component with a "burst-promoting" activity is present in anemic and normal mouse serum, precluding the requirement for "shaking".

Oddos et al.[15] further explored the idea of a component necessary for erythroid maturation in the mouse serum and showed that supernates from LTBMC which favored erythropoiesis contained a similar stimulating activity. By coupling frequent media changes with the addition of erythropoietin to the cultures, stimulation of erythropoietic maturation to day-2 CFU-E and erythrocytes occurred.

Our own experience is somewhat similar to that of Oddos et al.: By addition of erythropoietin and frequent media changes, erythroid progenitors, as late as 2-day CFU-E, can be maintained in these cultures. However, these progenitors can only be maintained for 4 weeks, and subsequently disappear from the cultures.

Culture Conditions Favorable to Erythropoiesis

Dishes. Petri dishes 60 mm in diameter were found to be better for the establishment of the adherent layer than 25 cm² tissue culture flasks, probably because adherence of the immediate precursors of CFU-E to plastic is lessened with Petri dishes, as described by others[13,15]. The contents of either one femur or tibia suspended in 5 ml complete media was plated per dish.

Medium. A complete medium based on Iscove's modified Dulbecco's MEM was chosen. This medium was found to be better buffered than α-MEM. Others[15] found have hyperacidification to be deleterious to the maintenance of erythropoiesis. The medium was supplemented with 15% horse serum,

206

5% fetal calf serum, 0.4 mg/ml human transferrin, 10^{-6}M hydrocortisone, 10^{-4}M β-mercaptoethanol, and penicillin-streptomycin. As has been the experience with standard LTBMC, horse serum must be pre-tested for its ability to establish an adherent layer in this particular culture system. Several lots of high-quality fetal calf serum that had been pre-tested for their ability to support growth of murine cell lines were found to be suitable.

Erythropoietin. After the initial marrow inoculum was placed in the dish, cultures were incubated at 37°C, 5% CO_2 to allow formation of the supporting stroma. For the next 2 weeks the cultures were disturbed only once to change the medium completely after 7 days. After 2 weeks each dish was recharged with 5 X 10^6 fresh, pooled marrow cells. At this point erythropoietin at a concentration of 0.5 U/dish was added to the complete medium, and remained as a supplement for the duration of the culture. We found that either murine or human erythropoietin produced adequate stimulation of CFU-E.

Feeding. The frequency of demidepopulation and refeeding of recharged cultures was most critical to success of the generation of CFU-E. The standard procedure of weekly demidepopulation with refeeding in LTBMC was inadequate to create favorable erythropoietin conditions. Only when we began demidepopulation and medium change twice or even thrice a week did we meet with success. The reason for this is probably to alleviate conditions that may lead to the exhaustion of the medium. Such conditions may include overacidification or possible generation of peroxides. This observation was first made by Oddos et al.[15] who coupled frequent changes of the medium with the addition of erythropoietin to circumvent the need for normal mouse serum in erythroid LTBMC. Although frequent handling of dishes contributes to contamination which plagues all LTBMC, it cannot be overemphasized that these frequent media changes were essential in order for the cultures to favor erythropoiesis.

Table 1. Total cell production and generation of primitive erythroid progenitor in long-term bone marrow cultures which favor erythropoiesis

	Non-Adherent Cells		Adherent Cells
days after Recharge	Cells/ml x10^4	Day 2 CFU-E colonies/10^5 cells	Day 2 CFU-E colonies/10^5 cells
day 2	28.0 ± 5.2	286 ± 25*	130 ± 23
day 7	9.0 ± 3.6	100 ± 18	78 ± 12
day 14	2.6 ± 1.2	143 ± 37	85 ± 13
day 21	4.0 ± 1.2	150 ± 40	73 ± 12
day 28	0.8 ± 0.2	45 ± 10	15 ± 9

* mean ± standard deviation

<u>Generation of erythroid precursors</u>. Following recharge, erythropoiesis in these cultures was detectable within 2-3 days. To assay erythroid progenitors, groups of 3 dishes were sacrificed, non-adherent cells were collected and total cell number determined. Adherent cells were then scraped and collected. CFU-E in both adherent and non-adherent populations were cultured in methylcellulose in the presence of 1 U/ml erythropoietin according to standard methods[16]. Day 2 CFU-E were counted after staining of the cultures with benzidine. The results are presented in Table 1. Cummulative total cell number in the non-adherent population increased until day 21 when it reached a plateau. However, CFU-E continued to be generated in the cultures for the entire 28 days as evidenced by the number of CFU-E detectable in the non-adherent as well as adherent layer. After 28 days the number of CFU-E began to decrease, and further detection of erythroid progenitors after this time was not possible.

The supernate of these cultures was subjected to cytocentrifugation and stained with benzidine as well as Wright-Giemsa. Mature non-nucleated as well as some nucleated red cells, which stained positively with benzidine, could be seen. However, because of the short duration of the cultures, it was not possible to ascertain their origin from either exogenously added or endogenously produced sources.

REFERENCES

1. McCulloch, E.A., L. Siminovitch, J.E. Till, E.S. Russell, and S.E. Bernstein. 1965. The cellular basis of the genetically determined hemopoietic defect in anemic mice of genotype Sl/sl^d. <u>Blood</u> 26: 399.
2. Tavassoli, M. 1975. Studies on hemopoietic microenvironment. <u>Exp. Hematol.</u> 3: 213.
3. Tavassoli, M., and W.H. Crosby. 1968. Transplantation of marrow to extramedullary sites. <u>Science</u> 161: 54.
4. Wolf, N.S., and J.J. Trentin. 1968. Hemopoietic colony studies. V. Effect of hemopoietic organ stroma on differentiation of pluripotent stem cells. <u>J. Exp. Med.</u> 127: 205.
5. Eaves, A.C., and W.R. Bruce. 1974. In vitro production of colony-stimulating activity. <u>Cell Tissue Kinet</u> 7: 19.
6. Ihle, J.N., J. Keller, S. Oroszlan, L.E. Henderson, T.D. Copeland, F. Fitch, M.B. Prystowsky, E. Goldwasser, J.W. Schrader, E. Palaszynski, M. Dy, and B. Lebel. 1983. Biological properties of homogenous interleukin 3. I. Demonstration of WEHI-3 growth factor activity, mast cell growth factor activity, and histamine-producing cell-stimulating factory activity. <u>J. Immunol.</u> 131: 282.
7. Stephenson, J.R., A.A. Axelrad, D.L. McLeod, and M.M. Shreeve. 1971. Induction of colonies of hemoglobin-synthesizing cells by erythropoietin in vitro. <u>Proc. Natl. Acad. Sci. USA</u> 68: 1542.
8. Allen, T.D., and T.M. Dexter. 1976. Cellular interrelationships during in vitro granulopoiesis. <u>Differentiation</u> 6: 191.
9. Dexter, T.M., T.D. Allen, and L.G. Lajtha. 1977. Conditions controlling the proliferation of hemopoietic stem cells in vitro. <u>J. Cell Physiol.</u> 91: 335.
10. Greenberger, J.S. 1978. Sensitivity of corticosteroid-dependent insulin-resistant lipogenesis in marrow preadipocytes of obese-diabetic (db/db) mice. <u>Nature</u> 275: 752.
11. Testa, N.G., and T.M. Dexter. 1977. Long-term production of erythroid precursor cells (BFU-E) in bone marrow cultures. <u>Differentiation</u> 9: 193.
12. Williams, N., H. Jackson, A.P.C. Sheridan, M.J. Murphy, A. Elste,

and M.A.S. Moore. 1978. Regulation of megakaryopoiesis in long-term murine bone marrow cultures. Blood 51: 245.

13. Eliason, J.F., N.G. Testa, and T.M. Dexter. 1979. Erythropoietin-stimulated erythropoiesis in long-term bone marrow culture. Nature 281: 382.

14. Dexter, T.M., N.G. Testa, T.D. Allen, T. Rutherford, and E. Scolnick. 1981. Molecular and cell biologic aspects of erythropoiesis in long-term bone marrow cultures. Blood 58: 699.

15. Oddos, T., S. Nacol-Lizard, and J.P. Blanchet. 1987. Erythropoiesis in murine long-term bone marrow cell cultures: Dependence on erythropoietin and endogenous production of an erythropoietin stimulating activity. J. Cell Physiol. 133: 72.

16. Iscove, N.N., F. Sieber, and K.H. Winterhalter. 1974. Erythroid colony formation in cultures of mouse and human bone marrow: Analysis of the requirements for erythropoietin by gel filtration and affinity chromatography on agarose-concanavalin A. J. Cell Physiol. 83: 309.

EXPRESSION OF HOMEOBOX GENES IN HUMAN ERYTHROLEUKEMIA CELLS

Wei-Fan Shen, Corey Largman, Patricia Lowney,
Frank M. Hack, H. Jeffrey Lawrence

Department of Medicine
University of California
Davis School of Medicine
Veterans Administration Medical Center
Martinez, CA 94553

ABSTRACT

Because homeobox-containing genes play a major role in embryogenesis
and tissue identity in Drosophila and because similar genes encode tissue-
specific transcription factors in mammalian cells, we hypothesized that
homeobox genes might plan a role in hematopoietic differentiation and
lineage commitment. We therefore surveyed a number of human leukemic
cell lines for expression of homeobox-containing genes by Northern gel
analysis with probes from the Hox 2 cluster of homeobox genes on chromosome
17. We observed transcripts for Hox 2.1, 2.2, 2.3 and 2.6 in the erythroid
line HEL and for Hox 2.3 and 2.6 in the erythroid line K562. Using homeobox-
-specific probes we confirmed that the transcripts visualized contained
the homeodomains for each gene as well as the flanking sequences. The
myeloid lines HL60, KG1 and U937 did not express specific transcripts
for any of the 4 genes studied. However, all these cell lines demonstrated
bands when probed at low stringency with certain Hox 2 probes, indicating
the expression of other homologous but as yet unidentified homeobox genes.
Expression of Hox 2.3 and 2.6 was seen in some T and B lymphoid cell
lines. Induction of differentiation in HEL cells resulted in complex
modulation of expression of the Hox 2 genes. We have therefore observed
erythroid-restricted expression of certain Hox 2 homeobox containing
genes in human erythroid cell lines and modulation of that expression
with differentiation, suggesting a role for these genes in the regulation
of hematopoiesis. Different homeobox genes appear to be expressed in
non-erythroid leukemic cell lines.

INTRODUCTION

Of the 40 or so genes which regulate embryogenesis in Drosophila,
25 share a common 180 bp motif called the homeobox, which encodes a 60
amino acid helix-turn-helix DNA-binding domain[1,2]. Clusters of homeobox-
containing genes are also present in mammalian genomes, where they are
expressed in spatially and temporally specific patterns during embryogenesis[3-6]. These genes appear to encode transcription factors which
regulate development[2,7]. Recently two homeobox genes have been identified
which regulate tissue-specific expression of other genes. The Pit 1
protein binds to the growth hormone gene enhancer in pituitary cells

and activates transcription[8]. Similarly oct-2 is a homeobox-containing protein which binds to the immunoglobulin enhancer in B lymphocytes[9]. Thus homeobox genes appear to play an important role in the regulation of mammalian development and determination of cell fate. We proposed that homeobox genes might play a similar role in the control of hematopoietic differentiation and lineage commitment. We therefore undertook a study of the expression of homeobox genes from the Hox 2 cluster on chromosome 17[4] in human leukemic cell lines expressing a variety of hematopoietic phenotypes.

MATERIALS AND METHODS

The 11 cell lines (2 erythroid, 3 myeloid, and 6 lymphoid - 4 T and 2 B cell) used in this study are outlined in Table 1. All cell lines were propagated in either alpha medium or RPMI with 10 % fetal calf serum at 37 degrees C in a 5% carbon dioxide incubator and batches of 0.5-1.0 x 10[9] cells were harvested during log phase growth for isolation of RNA.

Northern blot analysis

RNA was isolated by the guanidinium thiocyanate method[10] and poly A-enriched RNA was obtained by chromatography on oligo-dT cellulose[11]. The mRNA was electrophoresed in 1.5% agarose/formaldehyde gells, transferred to Gene Screen, and cross-linked to the filters by UV irradiation. Probes were labeled by the random primer method and used for hybridization in buffer consisting of 50% formamide, 5x Denhardt's solution, 1.0% SDS, 5% dextran sulfate, 0.1% sodium pyrophosphate, and 0.1 mg/ml salmon sperm DNA at 55°C. The initial probes used included three genomic homeobox-containing fragments from the human Hox 2 locus: Hu1, a 1.7 kb Hind III fragment representing Hox 2.1; Hu2, a 2.1 kb EcoRI fragment containing the Hox 2.2 homeobox; and Hu5, a 5 kb EcoRI fragment containing the Hox 2.6 homeobox[4,12] and a partial cDNA for Hox 2.3 which we cloned from U937 cDNA library[13]. Homeobox specific probes were synthesized with a 26 bp homeobox oligonucleotide primer (an anti-sense probe to nucleotides

Table 1. Human Cell Lines Studied

Cell Line	Source	Predominant Differentiation Pathway
HEL	Erythroleukemia	Erythroid (19)
K562	CML[1] -blast Crisis	Erythroid (20
HL60	Promyelocytic Leukemia	Granulocyte/Macrophage (21)
U937	Histiocytic Lymphoma	Macrophage (21)
KG1	Acute Myelogenous Leukemia	Macrophage (20)
UCD-PC1	Multiple Myeloma	B Lymphocyte (22)
Su	Multiple Myeloma	B Lymphocyte (23)
CCRF-CEM	Acute Lymphoblastic Leukemia	T Lymphocyte (24)
Molt-3	Acute Lymphoblastic Leukemia	T Lymphocyte (25)
Molt-4	Acute Lymphoblastic Leukemia	T Lymphocyte (25)
Mo	Hairy Cell Leukemia	T Lymphocyte (26)

[1]Chronic Myelogenous Leukemia

176-151 of the Hox 2.1 homeobox genomic sequence) on M13 single strand
templates, followed by appropriate endonuclease restriction, and isolation
of the single strand homeobox probes on 4% Nu-sieve gels. Homeobox specific
probes were hybridized in the same buffer at 50°C. Filters were washed
with a final buffer of 0.1x SSC, 0.1% SDS at 55°C. Probes were stripped
by heating filters at 95 °C in 0,1x SSC, 0.1% SDS. The actin probe was
a 2.1 kb fragment of the human beta actin gene. The relative ratios
of homeobox-containing mRNA to actin mRNA were determined by densitometric
scanning of autoradiographs, and changes in homeobox mRNA were calculated
by normalization of the actin signal.

Differentiation Experiments

Erythroid differentiation of HEL cells was induced by treatment
with cytosine arabinofuranoside (Ara-C) (10^{-7} M), and differentiation
was indicated by positive benzidine staining after 5-6 days. Monocytic
differentiation of leukemic cells was induced by treatment with phorbol
12-myristate 18-acetate (TPA) (10^{-7} M). Differentiation was indicated
by adherence of > 90% of the cells and development of positive staining
for non-specific esterase by 5 days. Granulocyte differentiation was
induced in HL60 cells by treatment with cis-retinoic acid 10^{-6} M; differentiation was assessed by morphology and by nitroblue tetrazolium (NBT)
reduction.

RESULTS

When Northern blots of mRNA from 5 non-lymphoid leukemic cell lines
were probed serially with the Hox 2 probes described in the Materials
and Methods, transcripts for Hox 2.2 were seen only in the erythroid
line HEL as a 1.4 kb band, while Hox 2.6 transcripts were seen in both
erythroid lines HEL and K562 as 2.1 kb bands (Figure 1 A,B). Hox 2.1

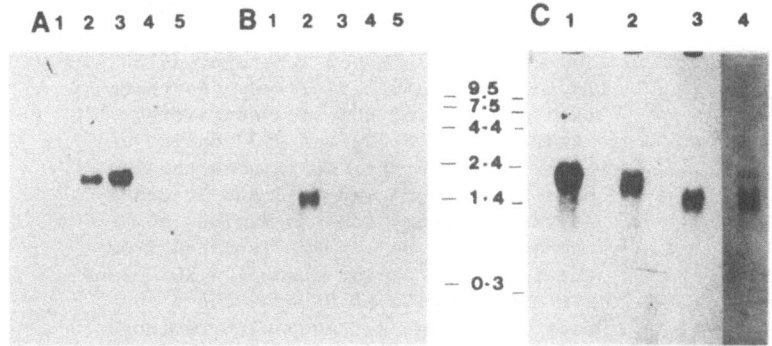

Fig. 1.   Northern blot analysis of mRNA from five human non-
          lymphoid leukemic cell lines. A. Hybridization of
          mRNA from KG1 (lane 1), HEL (lane 2), K562 (lane 3),
          HL60 (lane 4) and U937 (lane 5) with Hu5, a 5 kb Eco
          RI genomic fragment of Hox 2.6. B. The same filter
          as in A re-probed with Hu2, a 2.1 kb Eco RI genomic
          fragment of Hox 2.2. C. HEL mRNA probed with Hu5
          (lane 1), Hu1, a 1.7 kb Hind III fragment of Hox 2.1
          (lane 2), Hu2 (lane 3), and an Hel-8 cDNA probe, an
          800 bp Eco RI fragment representing the 5' flanking
          region of Hox 2.3 without the homeodomain (lane 4).
          Markers are RNA ladders (Bethesda Research Labs).

213

transcripts were visualized only in mRNA from HEL cells, while Hox 2.3 was expressed in both HEL and K562 cells (data not shown). Figure 1C shows HEL mRNA probed serially with all 4 Hox 2 probes under conditions of high stringenccy. Hox 2.1 transcripts are represented by a 1.7 kb band while Hox 2.3 appears as several bands centered about 1.4 kb. Reprobing the blots with homeobox-specific probes confirmed that the transcripts visualized contained homeodomains and not just flanking regions (data not shown). Probing with beta actin confirmed that RNA loading was approximately equivalent in all lanes (data not shown).

Since no transcripts were visualized in any of the non-erythroid cell lines when probed under highly stringent conditions, we probed another Northern blot of mRNA from the 3 myeloid lines KG-1, U937 and HL-60 with Hox 2.3 under somewhat reduced hybridization stringency (50 rather than 55 degrees). Figure 2 demonstrates that under these conditions a 2.2 kb transcript is seen in all three cell lines, while a larger 2.8 kb

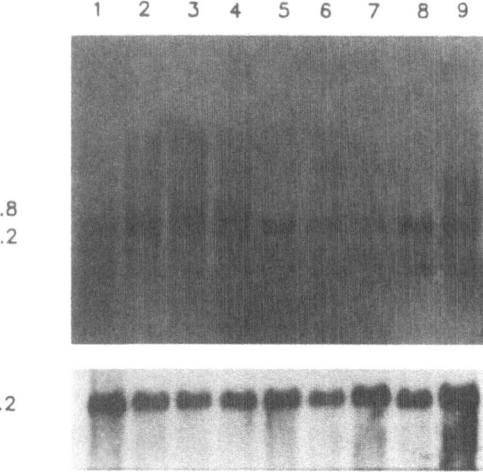

Fig. 2.   Non-specific homeobox gene expression
           in non-erythroid cell lines. Northern
           blot analysis of homeobox-containing
           mRNAs in KG1, HL60, and U937 cells. A
           Hox 2.3 cDNA probe (containing the
           homeobox region) was utilized at mod-
           erate stringency (50% formamide, 50
           degrees C) to probe mRNA isolated from
           the following cells: Lane 1 - KG1, lane
           2 - 4 U937 cells at 0 days, and 1 and 5
           days following treatment with retinoic
           acid; lanes; 5-9, HL60 at 0 days (lane 5),
           1 and 5 days following induction of granu-
           locytic differentiation with retinoic acid
           (lanes 6 and 7) induction of monocytic
           differentiation with 1,25 dihydroxy
           vitamin D3 (lane 8) and with phorbol
           ester (lane 9). A 2.2 kb band is seen
           in all the myeloid cells under all
           treatment conditions but appears dimin-
           ished following granulocytic differentiation
           of HL60. The 2.8 kb band appears re-
           stricted to U937 cells. The bottom panel
           shows the actin control.

band is seen for U937. Of interest is that the 2.2 kb band appears to be down-regulated in HL-60 cells treated with retinoic acid (lanes 6 and 7). When this blot was reprobed with Hox 2.3 under highly stringent conditions, no transcripts were visualized, strongly suggesting the expression of other as yet uncharacterized homeobox genes, with significant homology to Hox 2.3, in the myeloid cell lines.

To determine whether these Hox 2 genes were expressed in lymphoid cells, mRNA from 6 lymphoid leukemic cell lines was studied by Northern blot analysis. Figure 3 shows that probing with Hu1, a genomic probe for Hox 2.1 which contains both the homeodomain and 3' flanking region, revealed bands for all 6 lymphoid cell lines (bottom panel). However when the same blot was reprobed with a 3' flanking region Hox 2.1 probe which lacked the homeobox, bands were seen only in the positive control lanes containing HEL mRNA (upper panel, lanes 8 & 9). Thus, in a situation analogous to that seen in the myeloid lines, lymphoid cells appear to express homeobox genes whose homeodomains have significant homology with Hox 2.1. We have subsequently shown that Hox 2.2 is expressed in none of these 6 lymphoid lines, while Hox 2.3 and 2.6 are expressed in 1/4 and 2/4 T cell lines respectively and in both B cell lines (data not shown).

To determine whether expression of homeobox genes was modulated during the process of differentiation, we isolated mRNA from HEL cells before and after treatment with phorbol ester, which induces macrophage differentiation, and cytosine arabinoside, which induces erythroid differentiation, and performed Northern blot analysis with the 4 Hox 2 probes. As Figure 4 demonstrates, after correcting for differences in loading (see actin control bands), macrophage differentiation was associated with a transient reduction of expression of all 4 homeobox genes, especially Hox 2.1 and 2.3. In contrast, erythroid differentiation is associated with up-regulation of all four genes. Of note, probing mRNA from unfractionated normal human bone marrow revealed no transcripts for any of the 4 genes studied (lane 1).

DISCUSSION

Our study demonstrates that 1) certain members of the Hox 2 locus of homeobox genes on chromosome 17 are expressed in erythroid cell lines, 2) the transcripts contain homeodomains and not just flanking regions, and 3) expression of these genes is modulated during induction of differentiation. We have recently extended our survey of expression of homeobox genes to 18 human leukemic cell lines and observed that, with rare exception, Hox 2.1 and 2.2 are expressed only in erythroid cell lines[13].

It appears that other homologous homeobox genes are expressed in myeloid and lymphoid cell lines. We have recently cloned and sequence a partial cDNA for a novel homeobox-containing gene PL1, which is expressed only in cell lines with monocytic potential, such as U937[13]. Our data indicate therefore that certain homeobox genes are expressed in a lineage-restricted fashion, while others, such as Hox 2.3 and 2.6 are not. Studies of murine leukemic cell lines by Kongsuwan et al[14] have reported similar lineage restricted patterns of certain homeobox genes, although they observed that murine Hox 2.3 was expressed primarily in monocytic cell lines while we observed expression of human Hox 2.3 in both erythroid and lymphoid cell lines and not in monocytic cell lines.

Other malignant human cell lines have been shown to express homeobox-containing genes. In the human terato-carcinoma NT2, genes from the Hox 2 locus are activated by induction of differentiation with retinoic acid[4]. Similarly others have shown that, in a human colon cancer cell line, levels of homeobox mRNA increase during differentiation[15]. Our data are consistent with the hypothesis that mammalian homeobox-containing genes play a role in determination of cell phenotype and the regulation

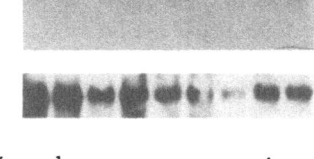

Fig. 3. Homeobox gene expression in lymphoid cell lines. Northern blot
analysis was performed on mRNA
isolated from four T cell leukemic
cell lines: CCRF-CEM (lane 1),
Molt 3 (lane 2), Molt 4 (lane 3),
and Mo (lane 4); as well as from
two B cell leukemic cell lines:
UCD-PC1 (lane 5), and Su (lane 6),
Control lanes include negative
control U937 (lane 7) and two
positive control HEL mRNA samples
(lanes 8 and 9). The top panel
shows the Northern blot following
hybridization under high stringency
conditions (55 degrees, 50% formamide) with a 350 bp 3' Hox 2.1
region probe lacking the homeodomain
and the poly A tail. This probe
hybridizes only to the positive
HEL control lanes. The middle panel
shows the same filter re-hybridized
to a Hox 2.1 genomic probe which
contains both the flanking regions
and the Hox 2.1 homeodomain. A 2.1
kb band is present in all lanes using
this probe. The bottom panel shows the
actin control.

of differentiation. Demonstration of expression of homeobox genes in
normal human marrow would strengthen this hypothesis in the case of hematopoiesis. Although we cannot detect homeobox transcripts in normal marrow
or peripheral blood leukocytes by Northern blot analysis, we have recently
been able to visualize Hox 2.2 in mRNA from normal human marrow by using
the more sensitive techniques of RNAase protection and polymerase chain
reaction[13]. These methodologies will permit us to study the modulation
of homeobox expression in normal bood cells.

The role of these genes in human leukemia remains very speculative.
Other investigators have recently demonstrated that the murine leukemia

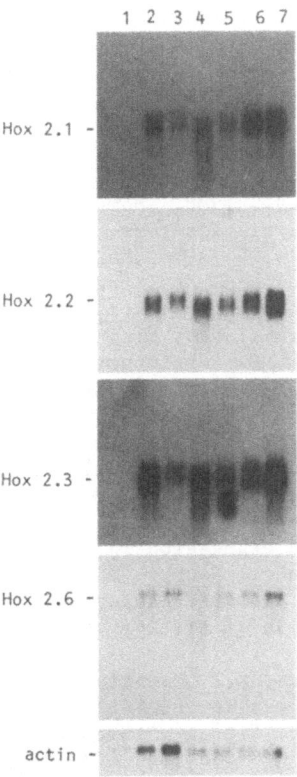

Fig. 4.   Changes in homeobox gene transcripts
          following HEL differentiation. North-
          ern blot analysis of HEL mRNA at day 0
          (lane 2), 6 hr (lane 3), 1 day (lane 4),
          and 5 days (lane 5) following induction
          of monocytic differentiation with
          phorbol ester, and 1 day (lane 6) and
          6 days (lane 7) following erythroid
          differentiation with cytosine arabino-
          side. The filter was sequentially probed
          with the respective homeobox-containing
          probes and with a beta actin probe. It
          should be noted that lane 1 contains
          mRNA isolated from a normal bone
          marrow. Although an actin signal is
          evident, no homeobox gene products were
          detected.

cell line WEHI 3 contains a proviral insert adjacent to Hox 2.4, which is expressed in significant levels only in this cell line[16,17]. In other murine myeloid leukemias, a deletion in chromosome 2 was associated with the deletion of a homeobox gene[18]. If homeobox-containing genes do play a role in regulation of hematopoietic differentiation, they may well be targets for leukemogenic events.

ACKNOWLEDGEMENTS

This work was supported by the Research Service of the Veterans Administration and the Leukemia Research Foundation. We are grateful to Dr. Craig Hauser for furnishing the Hox 2 probes and for many helpful discussions and to Dr. Thalia Papayannapoulou for furnishing the HEL cells.

REFERENCES

1. Gehring, W.J. 1987. Homeo boxes in the study of development. Science 236: 1245.
2. Levine, M., and T. Hoey. 1988. Homeobox proteins as sequence-specific transcription factors. Cell 55: 537.
3. Simeone, A., F. Mavilio, D. Acampora, et al. 1987. Two human homeobox genes, c1 and c8: Structure analysis and expression in embryonic development. Proc. Natl. Acad. Sci. USA 84: 4914.
4. Hauser, C.A., A.L. Joyner, R.D. Klein, T.K. Learned, G.R. Martin, and R. Tijan. 1985. Expression of homologous homeo-box-containing genes in differentiated human teratocarcinoma cells and mouse embryos. Cell 43: 19.
5. Simeone, A., F. Mavilio, L. Bottero, et al. 1986. A human homoeo box gene specifically expressed in spinal cord during embryonic development. Nature 320: 763.
6. Mavilio, F., A. Simeone, A. Giampaolo, et al. 1986. Differential and stage-related expression in embryonic tissues of a new human homeobox gene. Nature 324: 664.
7. Han, K., M.S. Levine, and J.L. Manley. 1989. Synergistic activation and repression of transcription by Drosophila homeobox proteins. Cell 56: 573.
8. Ingraham, H.A., R. Chen, H.J. Mangalam, H.P. Eisholtz, S. Flynn, et al. 1988. A tissue-specific transcription factor containing a homeodomain specifies a pituitary phenotype. Cell 55: 519.
9. Ko, H-S., P. Fast, W. McBride, and L.M. Staudt. 1988. A human protein specific for the immunoglobulin octamer DAN motif contains a functional homeobox domain. Cell 55: 135.
10. Chirgwin, J.M., E.R. Przyblyla, R.J. MacDonald, et al. 1979. Isolation of biologically active ribonucleic acid from sources enriched in ribonuclease. Biochem. 18: 5294.
11. Maniatis, T., E.R. Fritsch, and J. Sambrook. 1982. Molecular Cloning: A Laboratory Manual. Cold Spring Harbor Lab, NY.
12. Joyner, A.L., R.V. Lebo, Y.W. Kan, R. Tjian, D.R. Cox, and G.R. Martin. 1985. Comparative chromosome mapping of a conserved homeobox region in mouse and human. Nature 314: 173.
13. Shen, W-F, C. Largman, P. Lowney, J. Corral, C.A. Hauser, T.A. Simonitch, F.M. Hack, and H.J. Lawrence. Lineage-restricted expression of homeobox-containing genes in human hematopoietic cell lines, submitted.
14. Kongsuwan, K., E. Webb, P. Housiaux, et al. 1988. Expression of multiple homeobox genes within diverse mammalian hematopoietic lineages. EMBO Journal 7: 2131.
15. Sebastio, G., M. D'Esposito, M. Montanucci, A. Simeone, S.

Auricchio, and E. Boncinelli. 1987. Modulation expression of human homeobox genes in differentiating intestinal cells. Biochem. Biophys. Res. Comm. 146: 751.

16. Blatt, C., D. Aberdam, R. Schwartz, and L. Sachs. 1988. DNA re-arrangement of a homeobox gene in myeloid leukaemic cells. EMBO J. 7: 4283.

17. Kongsuwan, K., J. Allen, and J.M. Adams. 1989. Expression of Hox-2.4 homeobox gene directed by proviral insertion in a myeloid leukemia. Nucleic Acids Res. 17: 1881.

18. Blatt, C., and L. Sachs. 1988. Deletion of a homeobox gene in myeloid leukemias with a deletion in chromosome 2. Biochem. Biophys. Res. Comm. 156: 1265.

19. Martin, P., and T. Papayannapoulou. 1982. HEL cells: A new human erythroleukemia cell line with spontaneous and induced globin expression. Science 216: 1233.

20. Koeffler, H.P., and D.W. Golde. 1980. Human myeloid leukemia cell lines: A review. Blood 56: 344.

21. Harris, P., and P. Ralph. 1985. Human leukemic models of myelo-monocytic development: A review of the HL-60 and U937 cell lines. J. Leukocyte Biol. 37: 407.

22. Miller, C.H., A. Carbonell, R. Peng, T. Paglieroni, and A. MacKenzie. 1982. A human plasma cell line: Induction and characterization. Cancer 49: 2091.

23. Goldblum, R., A.S. Goldman, B. Rudloff, and N.S. Harris. 1973. An enhancing factor for in vitro growth of human plasma cells. Proc 7th Leukocyte Cultere Conference, pp. 15-28. Academic Press, New York.

24. Foley, G.E., H. Lazarus, S. Farber, B.G. Uzman, B.A. Boone, and R.E. McCarthy. 1965. Continuous culture of human lymphoblasts from peripheral blood of a child with acute leukemia. Cancer 18: 522.

25. Minowada, J., T. Onuma, and G.E. Moore. 1972. Rosette-forming human lymphoid cell lines. I. Establishment and evidence for origin of thymus-derived lymphocytes. J. Natl. Cancer Inst. 49: 891.

26. Saxon A., R.H. Stevens, and D.W. Golde. 1978. Lymphocyte variant of hairy-cell leukemia. Annal. Internal Med. 88: 323.